P U Z Z L E W I S

A simple step-by-step year-round program to augm... public, private, and home school students develop into knowledgeable, independent self-motivated life-long learners

GOALS

- Increase students' state mandated science test scores by 20%
- Increase parental involvement by 200%
- Increase school-home communication by 200%
- Increase enjoyable grade-level science activities outside of school by 30%

PUZZLE BUDDY CONTACT

Please write the name of your Puzzle Buddies and their relationships to you here:

Name: 1. ______________________ 2. ______________________
Relationship: 1. ______________________ 2. ______________________

PARENT CONTRACT

Dear Parents and Guardians,

This school year we request that you support your child's science learning by ensuring your child completes at least one crossword puzzle in this book each week. Every ______________________ your child should bring his/her puzzle book to school so the teacher can review the completed work and check progress. Please check the statements below that work for your family, and sign this contract showing your support for this program.

- ☐ I have read the Helpful Hints for parents, and will follow those that make sense for our family.
- ☐ I will make sure my child takes his/her puzzle book to school on the scheduled day.
- ☐ I will monitor my child's progress on a regular basis, encouraging him/her to keep up with the weekly schedule.
- ☐ I understand that my child does not have to finish the puzzle homework assignments. It's more important that my child tries hard, because success is often achieved with steady steps over time.
- ☐ I will consider buying a copy of this book for myself so I can improve my science skills and be more help for my child in the higher grades.
- ☐ I will notify my child's teacher if my child experiences repetitive frustration.
- ☐ I will help my child see that working on these crossword puzzles is fun and purposeful.
- ☐ I will help my child find a safe place in our home where this book can be kept. I will notify the teacher if my child loses or misplaces this PuzzleWise™ book.

I agree to the statements that I have checked above.

______________________ ______________
Parent/Guardian Signature Date

Track Assessment Progress

(7/10 = 70%)

5	/	=	%
10	/	=	%
15	/	=	%
20	/	=	%
25	/	=	%
30	/	=	%
35	/	=	%
40	/	=	%
45	/	=	%

Notes From The Authors

Notes to Teachers: Please review the ***"Guide for Teachers and Parents"*** before you begin working the puzzles and lessons. The teachers that created this book have a passion for science education. This book is NOT intended to teach concepts but rather to be used as a tool to build science vocabulary and review science content. The time commitment for using this book with your students is minimal but the rewards are enormous. The first few lessons require a lot of modeling and should be completed as a whole class activity, but from this point forward, the students complete the lessons as meaningful science homework. (We have found that the problem solving strategies and time spent up front will pay dividends later.) The time commitment is a half an hour each week to review the puzzles and clarify any misconceptions. Then, once a month, about an hour is needed for the students to take the assessment. The activity questions are formatted to resemble state assessments in science so students and parents will become familiar with the expectations. Effort was taken to have a variety of activities that span earth, life and physical science, thus spiraling the science curriculum. Our hope is that by helping students solve these puzzles we create life long problem solvers.

Notes to Students & Parents: We hope you enjoy working with our puzzles. We have included terms that will help students to understand science lessons and activities in the classroom by building a base vocabulary. In solving the crosswords: First, try to solve the clues that you know. Second, use the glossary to see if you can solve a few more by looking up key words from the clue. Third, go back and see if some of the letters added give you a hint at any unsolved clues.

Our Program Has Two Parts:

Crossword Puzzles: Every crossword has the following elements: 25-45 clues that span life, earth and physical content as well as general science vocabulary necessary to conduct investigations, inquire about systems, and apply content knowledge to solve problems. Use the glossary and any additional resources to solve the various clues:

- Sentences: Fill in the blanks with the appropriate word
- Definitions: Find the word that matches the definition
- Word Choice: Choose the word that answers the question or makes the statement true

Lesson Activities:

- ✓ Two Multiple Choice Questions
  - Use the process of elimination, the glossary and your knowledge to choose from the given choices
- ✓ One Extended Response
  - Diagram Labeling: Use the Glossary and/or other resources to identify parts of a system
  - New Investigations: See "Investigation" in the glossary for assistance in creating a prediction and procedure steps for the given investigation
  - System Analysis: Use the bullets to guide your analysis of a given system
  - Conclusions: Use the given data to draw a conclusion to an investigative question

Resources: The following publications are suggested student research materials.

ScienceSaurus	Wright Source Company
Scholastic Science Dictionary	Scholastic Inc.
Science Dictionary	Delta

Internet Resources: Rather than listing sites (that routinely disappear) we recommend teaching students to search for information on the Internet using search engines and share interesting findings with the rest of the class.

Glossary: The back of the book has pages with content information that will be useful in solving the puzzles.

How We Aligned this Book to National Science Standards:

Our science puzzle books are research-based on National Science Education Standards (NSES), the American Association for the Advancement of Science (AAAS) Benchmarks for Science Literacy, and the National Research Council (NRC) that clearly communicate age/grade appropriate benchmarks of what students should know and be able to do in science. Student understanding in science begins with active engagement in hands-on, minds-on learning. Students are then able to understand visual and symbolic representations of science concepts and processes through reading and writing. Background knowledge, rich learning experiences in science, spiraling science concepts and meaningful vocabulary help students develop scientific literacy.

Contributing Authors

Georgi Delgadillo
East Valley School District
Spokane, WA

Georgi Delgadillo is currently a TOSA (Teacher On Special Assignment) in the East Valley School District, Spokane, Washington, specializing in science curriculum issues and alignment. She is also a member of the **S**cience **A**ssessment **L**eadership **T**eam (SALT) that is under the direction of OSPI (Office of the Superintendent of Public Instruction) writing and evaluating science WASL testing materials since 2001. Georgi was a Bill and Melinda Gates Foundation grant recipient and has coordinated two Department of Ecology Grants. She is also a member of NASA's *Solar System Educator* team providing NASA staff development for the Washington, Idaho and Oregon region. Georgi is a member of Washington Science Teacher Association (WSTA) as a regional representative and part of the professional outreach team that has presented at numerous conferences. She is a reading specialist and has taught in grades 1-8. Georgi has worked with two different science museums across the country building programs that support state science standards.

Stewart "Andy" Anderson
Clovis Point Intermediate School
East Wenatchee, WA

Stewart "ANDY" Anderson is currently teaching computer skills to 6th and 7th graders, and he is the Math and Science Coordinator for the Eastmont School District. Andy has been teaching in the public school system since 1976. He has taught kindergarten through college level classes in his career. He was awarded the computer TLP (Teacher Leader Project) grant that is funded through the Bill and Melinda Gates Foundation in 1999. Andy has been a presenter at many state conferences. He is a science consultant for various school districts across the state. Being a member of SALT (Science Assessment Leadership Team) which is under the direction of OSPI (Office of the Superintendent of Public Instruction), Andy writes and evaluates science items that are used to assess students for the state-wide test (WASL). Andy has a Masters degree and has earned his administrative credentials. In his spare time, Andy teaches college science and math method classes for future teachers to help them prepare to teach science and math in the classroom.

Brian Teppner
Sierra Heights Elementary School
Renton, WA

Brian Teppner is an intermediate teacher in Renton, Washington. He has a Masters degree in integration of the arts into curriculum (with a focus on multiple intelligences), a Bachelors degree in elementary education, and came to teaching after ten years as a landscaper. Brian puts his horticulture background to work as a member of the grounds crew for the Seattle Mariners. Brian has received the "Ahead of the Curve" award from the Renton Chamber of Commerce for showing innovations in teaching. Brian is helping to create a vision for technology and science in the Renton school district. Brian is a member of NSTA, WSTA, SALT (Science Assessment Leadership Team), and has presented his fun-filled approach to science inquiry to districts across the state of Washington as well as at local and national science conventions. The clear message to every student, every day is "*... success is a matter of effort, pride and a belief in yourself.*" Brian teaches reading, writing and math through science and social studies and with the aid of technology and a great sense of humor. *"After using the math PuzzleWise books with my students, I knew science books would help teachers & students."*

Mary Bennett Moore
Jason Lee Elementary School
Richland, WA

Mary Bennett Moore, a National Board Certified teacher (MCGEN 2003), integrates math, technology, music, and language arts into her third grade students' science learning projects to ensure each student meets and exceeds grade level expectations in science. Mary received the 2005 Amgen Award for Science Teaching Excellence and the 2001 Presidential Award for Excellence in Mathematics and Science Teaching. She is a board member of the Washington Science Teachers Association (WSTA) and co-chairs the state level WSTA Science and Engineering Contest. Mary is also a member of the Washington Office of Superintendent of Public Instruction's Science Assessment Leadership Team, the Science Curriculum Instructional Frameworks team, and serves as a teacher trainer for Washington State Leadership and Assistance for Science Education Reform (LASER). In 1991 Mary received her BAEd from Eastern Washington University and received her Masters degree from Heritage College in 1995. She currently holds six Washington State teaching endorsements including Science 4-12 and Chemistry 4-12. Mary has also received several local and state teaching awards, presented at state and national conventions, and has successfully written and co-authored grants totaling over $245,000 to support teaching and learning.

PUZZLEWISE™

SCIENCE

Level 5

36 Weekly Lessons and 9 Monthly Assessments

Science Skill and Science Literacy Practice Aligned with National and State Science Learning Standards and Assessments

Dr. Daniel Levine and Matt W. Beck
Founders, Test Best International, Inc.™

May 2006

First Edition

TO ORDER, PLEASE CONTACT:
Puzzlewise™
P.O. Box 28312
Bellingham, WA 98228 U.S.A.
1-360-650-0671
Fax 1-866-683-3219
www.puzzlewise.com
daniellevine@puzzlewise.com
mattbeck@puzzlewise.com

Guide for Teachers and Parents

"Whether you believe you can do a thing or not, you are right."
Henry Ford, Automobile Industry Leader

Our Vision: Every child is a knowledgeable, independent, self-motivated life-long learner.

PRE-TEACHING ACTIVITIES:

- Read the Parent Contract, the Contributing Authors, the Helpful Hints, and Note From the Authors pages.
- Review this Guide for Teachers and Parents in its entirety.
- Make a copy of Lesson 1, and work it yourself so you get a feel for the activity.
- Review the entire book quickly so you know how it's laid out.
- Consider cutting out the last page of every student book so the assessment answer keys are restricted.
- Reminder: if any students require additional remediation, word lists are available from our Web site.

ANNUAL SCHEDULE:

The key to your students' success is this home practice program, which you initiate in your classroom. Your biggest time commitment is only in the first four classroom hours, during which you provide the foundation for your students' success in content skills and life-long learning.

Note: Any step can be expanded or diminished at the teacher's discretion!

1

STEP 1: Days 1 – 3: (Whole class activity):

GOALS:

1. **Students learn research skills.**
2. **Students develop critical-thinking skills and problem-solving strategies.**
3. **Students are introduced to science content grade level expectations.**

- This step introduces the genre of crossword puzzles, and allows you to teach researching, problem-solving, and critical thinking skills so your students can become knowledgeable independent learners.
- **Do not distribute the puzzle books**. (Wait for Step 2. Also, you have permission to photocopy the following:)
- We recommend making an overhead transparency of Lesson 1's crossword puzzle, but of the grid sections only. Also, photocopy Lesson 1's crossword clues and distribute to your students. Since students only have the clues, they'll have to watch the grid transparency on the overhead. They won't be distracted and you'll maintain their full attention as you teach.
- Start with any clue, and teach your students how to work a crossword puzzle, solve unfamiliar problems, and use resources to acquire solutions (textbooks, dictionaries, ruler…). Work the clues together, as a class.
- Use modeling to solve unfamiliar problems; it's the critical key to success! Show students how to solve problems using resources, words, numbers, and pictures.
- Remember to use these four steps when introducing a new concept to your students:
    1. Have a student read the problem aloud.
    2. Ask if anyone in the class can solve the problem. Provide adequate wait time.
    3. Have the student explain how they solved the problem, or teacher explains.
    4. Offer other examples of the same type of problem if needed.
- Some classes need to spend only one or two days on Step 1, and some need more. Also, if your students need more practice, use the Lesson 2 (and Lesson 3) crossword puzzle. You decide when to advance your class to Step 2.

2 STEP 2: Days 4 – 6: (In-class independent or small group activity):

GOALS:

1. Students effectively communicate their understanding of science.

2. Students share strategies with their peers.

3. Students work toward independence such as reasoning logically and making connections.

- This step helps students achieve success at working toward independence by doing the same lessons (Lessons 1-3) individually or in pairs that they previously did as part of a large group.
- Dedicate 30 minutes a day of class time.
- Distribute the books. (Collect them at the end of each class.)
- Have students write both their name and their teacher's name on the inside front cover.
- Have students work on Lesson 1 on their own or in small groups. As this is a lesson they've already done as a class activity, it should be familiar and proceed easily.
- Stop work periodically to review, as a class, 2-3 problems with which students need assistance. This is an opportunity for a great teaching moment!
- Follow the four "new concept" steps, listed above in Step 1.
- After your students have done Lesson 1, advance to Lesson 2, etc., doing the lessons individually or in small groups. Some classes need to spend only a few days on Step 2, and some need more. You decide when to advance your class to Step 3.

  REMEMBER: Any step can be expanded or diminished at the teacher's discretion!

3 STEP 3: Remainder Of The School Year (Weekly home assignments):

GOALS:

1. Students enlist the support of a science helper (Puzzle Buddy).

2. Students practice and mature their scholastic independence.

3. Students build their knowledge base by steadily visiting/revisiting concepts.

4. The teacher's instruction is enhanced by spiral learning of the five content strands.

- Students are now ready to be independent learners, with emphasis on work-at-home. Students can take their books home!
- Assign students to identify 1-2 people outside school to be their "Puzzle Buddy." A Puzzle Buddy can be a parent, guardian, grandparent, other older relative, friend or neighbor. This person is a resource to assist them with clues.
- Students write the name(s) and relationship of their Puzzle Buddy on the inside front cover. During the year, students are encouraged to do their lessons with their Puzzle Buddy as a fun way of learning, and solving challenging problems. Their Puzzle Buddy becomes the main resource person.
- This is where the 36 weekly assignments begin. Students are to complete one lesson a week, on their own. Students write the due date in the space at the bottom of each page. Use your planning book to remind students about the lessons to be completed each week, and to remind yourself about the monthly assessment.

- Students bring in their books once a week for a 30-minute in-class activity time ("puzzle time") so the teacher can observe student progress and facilitate student peer-to-peer communication, critical thinking, and problem-solving strategies.
- Review the homework assignment in class. At the end of class, have students fill in the blanks even if they didn't get the answer themselves. (This increases engagement.)
- During long holidays, students must still complete one lesson a week. Skill maintenance and development is a year-round task!
- Also, now is the time to get parents involved:
  - o Have students ask their parent/guardian to read and sign the Parent Contract on the first page of this book.
  - o Throughout the year, students have their parent/guardian review the completed assignments, and initial the bottom of each page.
  - o Let parents know that in the beginning of the school year their child will be exposed to unfamiliar material, and it's all right if assignments are not completed. (Here's where the Puzzle Buddy can be a big help!) As the year progresses, more and more of the assignments will be completed. Most important is that their child tries hard, and develops and matures their critical thinking and problem solving strategies. Success is often achieved with steady steps over time.
- On assessment days:
  - o Once a month you may conduct an assessment using every fifth lesson. Allow enough time for students to stretch themselves as they work through their challenges.
  - o Conduct assessments on a different day than your weekly puzzle time.
  - o You may want to snip out the keys on the last "scissor" page.
  - o On the days your students do the assessment lessons (every fifth lesson), you may collect all books and record the results on the Parent Contract page (page one).
  - o Review the preceding lessons quickly to assure yourself about the quality of your students' work at home, and return the books the following day.
  - o Check for parent/guardian initials on each lesson page, and confirm that there is Puzzle Buddy information on the inside front cover.
- Additional assessment ideas:
  - o Daily Puzzle Clues (D.P.C.)

    Give the students 1-2 problems daily from their weekly assigned puzzle. Similar to D.O.L. and excellent for teaching moments.
  - o Weekly Puzzle Quiz

    Give your students a 5-10 question quiz from the assigned weekly puzzle. After they finish, have them pass the paper to another student and correct together as a class. You may have the students turn the papers in and monitor their progress.

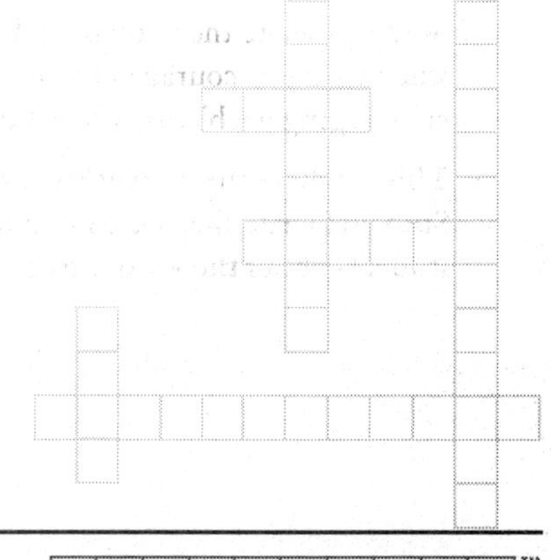

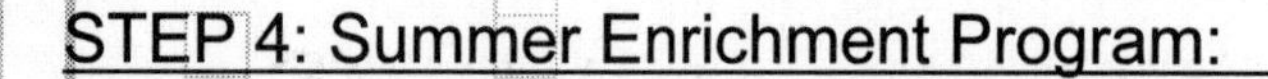

STEP 4: Summer Enrichment Program:

GOALS:

1. **Students become year-round learners.**
2. **Students maintain learned skills.**
3. **Students are introduced to some new skills required in the next grade.**

- If you have 15 or more lessons remaining at the end of the school year, these books make an excellent summer transition program. Finally, a summer math program everyone can do!
- Using the book this way helps students maintain the grade level skills they've just acquired, and introduces some of the skills they'll need to be effective, receptive learners in their new grade. Your students will look 'science-smart' with their new teacher!
- Teachers will want to communicate with their students' parents about how important it is that students continue to work and maintain their skills during the summer. The lessons at the end of the summer program book have a blend of the grade skills students have learned, as well as an introduction to the next grade's skills, offering students the chance to excel in the next grade!
- Remind students to turn in their books to their new teacher when school resumes.
- You may use this book as the summer transition program if you have 15 or more lessons remaining. In addition, teachers have been so successful using our books for a summer transition program that we are currently developing a Summer Enrichment Program book series. As soon as our Summer Program is ready, you'll be able to order from our on-line store at puzzlewise.com.

Summer Incentives and Contact Ideas:

Students like to be rewarded for doing a great job, so plan something fun for your students so they know this work is important, that you'll be checking their progress, and that there's a nice surprise waiting for them at the end. Some principals and teachers plan a small party for all the students that completed their book by summer's end; other schools call each family twice during the summer to cheer the students on; some schools have a mid-summer Puzzle Festival. This additional effort will result in huge gains in student development, and will also result in a much more effective teaching experience in the Fall when students spend less time re-learning the skills they were supposed to retain!

Helpful Hints to Ensure Student Success: Teachers

Genre of Crossword Puzzles

As you begin using the crossword puzzles with your students, it's important to help your students understand the genre of crossword puzzles. For example, students may wonder why there is a "1 across" clue and a "3 across" clue – but no "2 across." They also need to learn the strategies of completing a crossword puzzle (e.g. use the answers within other clues to figure out the answer for an unknown clue, and first answer the questions you know and later do the questions you don't know).

The best way to teach your students the genre of crossword puzzles is to follow the "gradual release of responsibility" model (Pearson and Gallagher 1983[1]), explicitly presenting crossword strategies and then gradually handing over responsibility while modeling and guiding.

First, model the strategies used in figuring out a crossword puzzle. Refer to the *Guide for Teachers and Parents* for introducing new concepts. Demonstrate the reading, writing, and mathematical problem solving as you "think aloud". Demonstrate how you use classroom tools (e.g. glossaries, dictionaries, science books, calculators, rulers) and prior knowledge to solve the puzzles. Model what you do when you come across a clue you can't solve. Model many times – one demonstration isn't enough. It's better to have many short demonstrations than one long demonstration. Next, begin solving puzzles in a large group setting (Step 1 of Guide). This is a shared endeavor! Participate with your students in problem solving and completing the puzzle; you lead and support as they do the work. Try to get the children to tell you what to write on the puzzle. Encourage students to share their thinking behind their recommendations.

When students are capable of further responsibility, break the students into small groups and step back from direct participation. Monitor your students and then assess whether you need to do further demonstrations. Pay attention to the challenges they are facing. Use this information to plan further demonstrations, which don't need to be done just by you – recruit some of your students who are excelling at solving crossword puzzles. Be sure the students "think aloud" as they share their answers and strategies.

Finally, when you feel your students have a strong sense of how to solve crosswords, have your students work independently. First, have them work independently in the classroom (Step 2 of Guide), and then, after some repeated successes, have students practice their crossword skills as homework (Step 3 of Guide). Continually assess your students as they work on their crossword skills. Ask questions to learn what strategies and mathematical skills they're using.

Value of Crossword Puzzles

Not only is it important that your students know HOW to figure out crossword puzzles, but it's also important they know WHY someone would want to do them. Neglecting this aspect could result in lower motivation. Here are some ideas to help your students understand the value of crossword puzzles:

- **Challenge your students to see where they can find crossword puzzles**; create a class list and maybe even post examples on a bulletin board. Students are likely to bring in examples from restaurant place mats, cereal boxes, TV guides, newspapers, magazines, etc. As students contribute to the list, ask them to predict the reasons the authors wanted to publish a crossword puzzle.
- **Have your students create a survey to find out some statistics of who completes crossword**

1 *Pearson, P.D. & Gallagher, M.C., (1983). The instruction of reading comprehension. Contemporary Educational Psychology, 8: 317-344.*

puzzles, how often, and the origin of the puzzles (e.g. magazines, newspapers, books, etc.). Students can survey family members, faculty members and students in the school as well as community groups such as firefighters, police officers, Rotary Clubs and senior citizen centers. Work this into your curriculum by having students graph the results of the survey. Do you see a trend with age groups? Gender? Why do they think that's so? Part of the survey should include a section where people share why they like crossword puzzles (if they do). For example, students may learn that people like the intellectual exercise provided by crossword puzzles and that puzzles fill up wasted time when a person is waiting for a bus or a doctor's appointment.

- Cruciverbalist: If you encounter the clue of "a 14 letter word that means crossword puzzle fan," this is the answer. Do some internet research and find out about crossword puzzle tournaments. Ask your students why people participate in these competitions. Maybe your class wants to create a class or school wide crossword puzzle tournament.
- Invite your students' family (including grandparents) to come into the classroom and sit on a "crossword puzzle panel" where students ask them why, when and how they solve crossword puzzles. When you have these people in the room, emphasize how these people rarely complete a crossword puzzle. Make the comparison of a famous sports figure – just as they do not play the game perfectly, it's not expected that the students complete every puzzle perfectly. Make sure the value is placed on the problem solving and sharing of strategies – not on the "right answer."
- Get your students' families to support this work! Share with your students' parents how these particular crossword puzzles have many benefits – such as they help prepare students for state standardized tests. Don't overemphasize the test preparation – be sure you communicate the many other benefits of crosswords as well.

Working Crossword Puzzles into Your Class Schedule

As described in our Guide for Teachers and Parents, after introducing the crossword puzzles to your class, you may want your students to complete a crossword puzzle each week as homework and then complete one in class per month as an assessment. There are a variety of strategies when using these crossword puzzles in the classroom. Here are some suggestions:

1. Work the puzzles into your daily science time.
2. Have a daily puzzle time (just as many classrooms have a daily silent reading time).
3. Have students work on their puzzles as "bell work" at the beginning of class as a transition to academics.
4. Place the students into puzzle teams where they work together on a regular basis to solve the puzzles.
5. When students complete their science assignments, make this one of their choice assignments. (Consider offering extra credit for puzzle work.)
6. Rather than doing the traditional "daily computation problems" that many classrooms do, assign crossword puzzle clues instead.

Remember, it's critical that students communicate about their scientific problem solving and crossword puzzle strategies. Many state tests require students to share their thinking, so it's important that students get in the habit of doing so – both verbally and in writing. Crossword puzzles can foster scientific communication as students explain and justify their answers to one another.

Be sure you're using the crossword puzzles as a tool to encourage more science communication within your classroom. By sharing their science thoughts with classmates, students have the opportunity to see the perspectives and methods of others, leading to expanded mental agility and increasingly creative strategizing.

Helpful Hints to Ensure Student Success: Parents

Recommendations for Parents and Guardians

Whether you are a home schooling parent or one who just wants to encourage additional science skill practice, read through the Helpful Hints for classroom teachers. Follow the guidelines where it makes sense to do so. As stated in the teachers' directions, it's critical that students communicate their strategies and problem solving.

If your child doesn't have another child with whom to communicate – then here are some great options:

1. **Communicate** with your child about the strategies and problem solving involved in completing the crosswords .
2. **Partner with other parents and their children**, allowing the children time to share their answers along with their reasoning.
3. **Partner your child with a senior** who enjoys crosswords so they can work them together.
4. **Start a puzzle club**. Have your child send out invitations to join the club that could meet weekly, biweekly or monthly. Include crosswords and other puzzles such as word jumbles and word searches.

Here are some hints to help your child complete the crossword puzzles in this book:

1. **Carry the puzzle book in the car** so it's available during car rides, waiting room visits, and other "I'm bored" times.
2. **Alternate completing answers**. Your child may answer the "across" clues and you may answer the "down" clues. Take turns completing answers.
3. Use the crossword puzzles as a **form of assessment** to help you plan the science instruction for the day. For example, if your child struggles with the puzzle clue, "Photosynthesis occurs in the plant's ______," then focus your daily science lesson on the life cycle of plants.
4. It's very important to understand that in the beginning of the school year your child is being exposed to unfamiliar material, and it's **all right if your child doesn't finish all the problems**.

As long as your child makes a good attempt to finish the assignment every week, that's sufficient. As the year advances, and more skills are learned and matured, more of each puzzle will be completed. It's more important that your child tries hard, because success is often achieved with steady steps over time.

5. Remember, if you want your child to embrace crossword puzzles as a leisure time activity, you need to model working on crossword puzzles **during your "off time,"** too! When you show yourself as a life long learner who enjoys puzzles, your child is likely to do the same.
6. Make sure there's **laughter** worked into your puzzle time, so crosswords are not "a chore." Do all you can to make the puzzles enjoyable, appealing and worthwhile to your child!

ROXANN ROSE (*Author: Teacher and Parent Guides*) is an innovative teacher, national presenter, and respected educational consultant. Ms. Rose taught elementary students in three different states, on both coasts of the United States for 11 years. Roxann's teaching is highlighted in Harvard Project Zero's Creative Teaching Video Series. She continues to work with Disney Learning Partnership to recognize stellar teachers across the United States. Roxann teaches future teachers and works with current teachers to help them reflect on and improve their teaching practices.

The PuzzleWise™ Foundation

Like you, we believe in service to others. There are great needs of all kinds in our families, our communities, and our world. We want to be part of the solution, and we've established a non-profit foundation to address some of these needs both locally and globally. A portion of sales will be given to the foundation for the growth and well-being of those in need.

One of our initial ideas is to establish a network of organic food restaurants in communities across America where families and people can receive a healthy, tasty dinner at cost. Too often our friends and neighbors come home after a hard day's work, and there's no time to prepare a healthy meal for themselves and their children. We rely too much on processed foods loaded with impurities, and the foods we eat further compound the personal and social problems we face today.

Instead, families and individuals can have a balanced dinner rich in vitamins and minerals, which tastes great and empowers their bodies, and only pay the actual cost. We further envision these restaurants becoming a place where families gather not only for food, but also for social connection and for rebuilding the relationships of a healthy community.

We want to establish a model that serves 100% of the families in our community. Once people come to the restaurant to eat healthy food at low cost, we want the community to find ways to help its members. Elders may need their lawns mowed or firewood stacked, a family might need help moving furniture or setting up a new business, perhaps the local youth can be organized to help in the park, maybe a shelter needs to be established, someone needs medical or dental attention, someone's car needs a little work, someone has a job to offer a person who needs one... We need to rebuild our community, and there's no one to do it but us!

Yes, this is an ambitious program, and we'll do what we can. To accomplish the possible, we all have to work together and put forward our best attitudes to match our strongest attributes. And we can't do it alone; we have to do it together, as a **TEAM**: **T**ogether **E**verybody **A**chieves **M**ore.

"Anyone can count the seeds in an apple. No one can count the apples in a seed." (Anonymous)

We hope you'll like our books and support our common cause. Thank you!

With our best wishes,

Matt and Daniel

New Books From PuzzleWise™

Math: Levels 1-8:

"I am convinced these books have a positive impact on my students' reasoning, problem-solving abilities and mathematics knowledge. I know this program is helping me prepare my students for their future."
—Jane Robertson, Arizona Teacher of the Year, 2004.

Innovative Ways to Use Our Books:
We have several new options to help you incorporate our books into your teaching program, in the way best suited to helping *your* students.

For additional remediation, we offer word lists for every puzzle on our Web site. You choose when to offer word lists, depending on the needs of your students.

You can also use our books in a **summer enrichment program.** If you plan your schedule so that 15 or more puzzles remain to be completed over the summer, our math books can double as your summer transition materials!

Visit

www.PuzzleWise.com

- Download free sample puzzles!
- Order books!

Find an educational resource store near you!
Please contact the educational resource stores listed at www.puzzlewise.com to purchase any of these excellent teaching materials locally. You'll save the cost of shipping, and help your local economy.

Thank you!

Dr. Daniel Levine
1-360-650-0671
Fax 1-866-683-3219
daniellevine@puzzlewise.com

Lesson 1: "If it is to be, it is up to me."

Anonymous

/ 25 = %

ACROSS

4 This shows us evidence of life long ago.
6 The force that attracts metals to other metals.
13 The form of energy that changes liquid water into a gas. (2 wds.)
14 A natural structure on the surface of the Earth. (2 wds.) (l____ f____)
16 Measured in inches, feet, and yards. (d____)
17 Water vapor is this state of matter.
19 A curved piece of clear glass that changes the direction of (bends) light.
21 This is what people eat.
22 The flash of light caused by particles moving from one cloud to another and strike the Earth.
23 This is the smallest unit of life.
24 The gas surrounding Earth.
25 Igneous and sedimentary rocks changing to metamorphic rocks is an example of the rock _____.

DOWN

1 This is the first part of the digestive system.
2 _____ are a part of the skeletal system.
3 The amount of matter in an object.
5 Another word for valid, clear, reasonable.
7 The tube that absorbs nutrients in the digestive system.
8 A solid natural material of only one substance.
9 The heart, liver, and lungs are examples of an _____.
10 Cord like cells that connect our sense organs to our brain.
11 Organisms that eat dead material and waste are called _____.
12 A (consumer producer) eats plants or animals.
15 Proof to show something is true.
18 A general term for deer, insects, humans. (a____)
20 The food chain shows this moving through the ecosystem.

Parent/Guardian Signature: For puzzle time, return on:

1. How do butterflies help plants?

O A. Butterflies drink flower nectar.

O B. Butterflies pollinate flowers.

O C. Butterflies lay eggs on plants.

2. How do caterpillars harm plants?

O A. Caterpillars climb on plants.

O B. Caterpillars leave waste on plants.

O C. Caterpillars eat plant parts.

3. Draw and label the 4 stages in the life cycle of a butterfly.

- Include these labels in your drawing:
  - ✓ Adult
  - ✓ Larva
  - ✓ Egg
  - ✓ Chrysalis

Use words, labeled pictures, and/or labeled diagrams in your answer.

Draw and label a butterfly life cycle below:

Lesson 2: "Good enough never is."
Debbi Field, Founder of Mrs. Field's Cookies

/ 25 = %

ACROSS

6 Plant systems get their energy from the _____.
7 Describing how high a sound is.
9 Material that has never been alive. (n_____)
11 Warm-blooded animals are called _____.
15 Another word for precipitation.
17 _____ is a property of an object.
18 An animal that preys on other animals.
20 A _____ can be as small as an insect or as large as the universe.
22 Plants need carbon dioxide, sunlight, and _____ for photosynthesis.
23 Land surrounded by water on all sides.
24 Melted rock that comes out of a volcano.

DOWN

1 Outcome of an investigation.
2 Some plants such as trees and grasses need _____ for pollination. (snow rain wind)
3 Earth's natural satellite.
4 A gas produced by plants needed for animal survival.
5 A simple machine that transfers energy.
6 Another name for dirt.
8 Measured in days, months, and years. (rhymes with nine)
10 The energy produced by a machine. (output inputs)
12 Metric measurement of distance.
13 Form of an object. (rhymes with cape)
14 Uses the process of photosynthesis to create energy. (producer consumer)
16 An organism that is eaten by predators for food.
19 Melted rock below the Earth's surface.
21 Energy from the sun. (s_____)

Parent/Guardian Signature: For puzzle time, return on:

1. An insect has eaten all of the leaves off of a plant. How would this affect the plant system's ability to survive?

O A. Losing its leaves would not affect the plant system's ability to survive.
O B. The plant would not be able to make its own food through photosynthesis.
O C. The plant would not be useful to other organism as a source of food.

2. What makes a plant a producer?

O A. Plants are able to make their own food.

O B. Plants are able to grow and reproduce.

O C. Plants are able to make fruit and seeds.

3. What impact do animals have on plants?

- Identify **one** way that animals **help** plants.
- Identify **one** way that animals **harm** plants.

Use words, labeled pictures, and/or labeled diagrams in your answer.

One way animals help plants:

One way animals harm plants:

Lesson 3: "If you stand at all, stand tall."
King James I of England

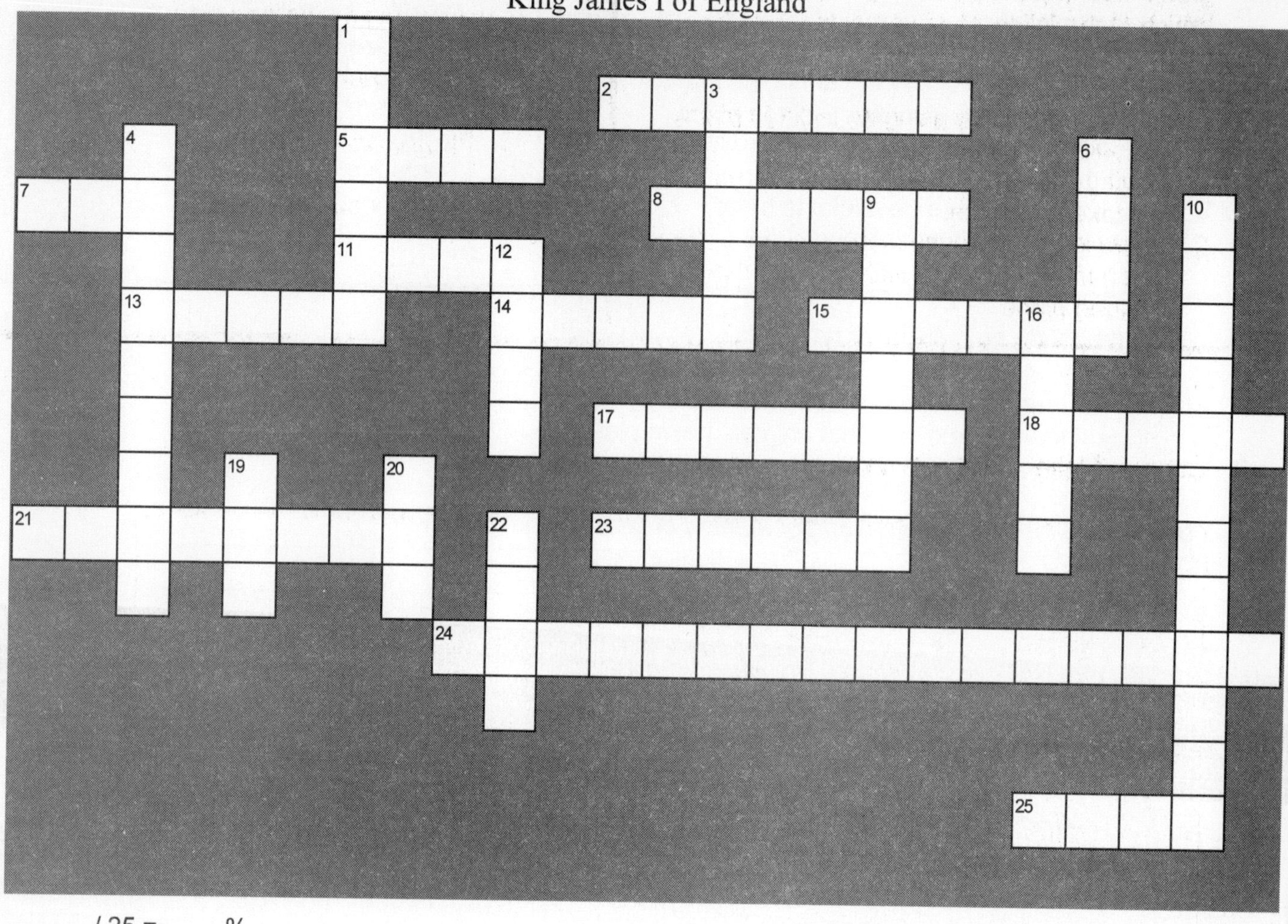

/ 25 = %

ACROSS

2 This is where a plant survives best. (h____)

5 Earth material made of tiny rock pieces, minerals, and plant matter.

7 Mixture of gases that surrounds the Earth.

8 Paperclips, nails and iron filings are attracted to a _____.

11 The part of the plant where food is produced through a process called photosynthesis.

13 The life _____. (c____)

14 Salt water that covers much of the Earth.

15 The dusty powder that is transported by insects from plant to plant.

17 When you receive traits from your parents. (rhymes with parrot)

18 A large stream.

21 During an investigation only one _____ is changed.

23 Plants produce a gas that humans and animals need.

24 _____ are absorbed by roots. (2 wds.) (m____ n____)

25 _____ is a liquid form of precipitation.

DOWN

1 An elastic tissue that helps animals move.

3 The organ that allows humans to process information.

4 Steps to a scientific investigation.

6 Steps to complete an experiment. (rhymes with man)

9 The wearing away of soil and rock is called _____.

10 The process by which water changes from a gas form back into a liquid. (perspiration condensation)

12 We eat _____ to give us energy.

16 Our planet the _____, has many different weather patterns.

19 An expanding substance without a fixed shape. (one of the 3 states of matter)

20 A body of salt water smaller than an ocean.

22 Water in a solid form that falls from a cloud.

Parent/Guardian Signature:

For puzzle time, return on:

1. Andy needs to lift a heavy box that is on the ground and put the box in the back of a truck. Which of the following statements would be true?

O A. Moving the box by using an inclined plane will take more force.
O B. Moving the box by using an inclined plane will take less force.
O C. There is no difference in the amount of force when moving the box with or without an inclined plane.

2. Andy wondered how people use an inclined plane. Which of the following is an example of a person using an inclined plane?

O A. Using a ramp to load a truck.

O B. Playing on a teeter-totter.

O C. Driving a car.

3. Draw and label a diagram using an inclined plane in every day life:

Lesson 4: "Problems are only opportunities in work clothes."
Henry Kaiser, Builder

/ 25 = %

ACROSS

3 The sprouting of a seed.
7 The natural satellite seen from Earth.
10 The color of an object. (p____)
12 _____ is a visible property of an object. (c____)
13 The sun provides this for plants.
18 Humans breathe this gas to function.
19 A (lever screw wedge) is an incline plane wrapped around a center pole.
20 Melted rock inside of a volcano.
21 The items used in a science demonstration. (arterial material)
22 Organisms that eat both plants and animals are called _____.
23 The only star in our solar system.
24 A large object that orbits a star.

DOWN

1 A large body of water surrounded by land.
2 Plants can make their own _____.
4 When water changes from a liquid to a gas. (evaporate examinate)
5 12 of these make a foot.
6 To describe an outcome. (e____)
8 Referring to the height of an object. (s____)
9 The smallest part of an organism. (rhymes with bell)
11 The process by which animals create new offspring. (rhymes with reconstruction)
12 A plant growing from a seed to make new seeds is an example of the life _____.
14 Measured in centimeters, meters, and kilometers. (d____)
15 This part of a plant system is where the plant makes its own food.
16 A dark funnel of strong winds that spiral upwards.
17 An organism that moves on its own and eat food.

1. How do mineral nutrients in the soil benefit a plant?

O A. They are absorbed by roots with water.

O B. They provide a footing for roots to hold onto.

O C. They help a plant to grow and make seeds.

2. Which variable is changed if one plant is placed in a window, one in a closet, and one in an open area in the same room?

O A. Amount of light

O B. Amount of water

O C. Amount of soil

3. What are two things plants need to grow and be healthy?

- Identify **two** needs of a plant system to survive.
- Explain why the plant needs these things to survive.

Use words, labeled pictures, and/or labeled diagrams in your answer.

First Need:

Why do plants need this?

Second Need:

Why do plants need this?

Name:

/ 25 = %

ACROSS

6 This is what people eat.
7 Deer, insects, humans for example.
8 Measured in inches, feet, and yards.
9 The smallest part of an organism.
11 12 of these make a foot.
16 During an investigation only one ______ is changed.
17 An animal that preys on other animals.
18 Melted rock below the Earth's surface.
20 The natural satellite seen from Earth.
22 Material that has never been alive.
23 Water vapor is this state of matter.
25 Salt water that covers much of the Earth.

DOWN

1 Another name for dirt.
2 The process by which water changes from a gas form back into a liquid.
3 Referring to the height of an object. (s_____)
4 Water in a solid form that falls from a cloud.
5 This is where a plant survives best. (h_____)
10 Melted rock that comes out of a volcano.
12 The sun provides this for plants.
13 Heart, liver, and lungs are examples of ______.
14 The sprouting of the seed.
15 An organism that is eaten for food.
19 The amount of matter in an object.
21 Plants produce a gas that humans and animals need.
24 Plant systems get their energy from the ______.

Parent/Guardian Review: Date:

1. What three things do plants with leaves need in order to make their own food?

O A. Sunlight, carbon dioxide, water

O B. Water, sunlight, soil

O C. Oxygen, water, sunlight

2. Which variable is changed if one group of plants is placed in a 72° F environment and one group of plants is placed in a 62° F environment?

O A. Type of plants

O B. Amount of light

O C. Temperature

3. A plant is a system made up of parts that work together.

- Draw and label a plant system. Include these labels in your drawing:
  - ✓ roots
  - ✓ stem
  - ✓ leaves
  - ✓ flowers
- Explain what would likely happen if one of the parts of a plant system were missing.

Use words, labeled pictures, and/or labeled diagrams in your answer.

Draw and label a plant system below:

Missing plant part:
How will this missing part affect the plant system?

Lesson 6: "There's a better way to do it. Find it!"

Thomas Edison, Inventor

1 2 3 4 5 6 7 8 9 10 11 12 13 14 15 16 17 18 19 20 21 22 23 24

/ 25 = %

ACROSS

1 Light, water and _____ are important for plants to survive. (2 wds.) (m_____ n_____)
5 Share the results of an investigation. (rhymes with support)
7 The first leaves to grow are called seed leaves or _____. (c_____)
8 The source of energy for the water cycle.
11 The process of getting food from one organism to another. (2 wds.) (f_____ c_____)
14 Scientific term to explain a plant that is beginning to grow.
16 Visible traits. (rhymes with idealistic)
20 To make a guess about the outcome of an event. (p___)
21 A _____ is a producer.
23 An imaginary line that passes through the Earth's north pole to the south pole.
24 A light bulb glowing is a form of _____. (2 wds.)

DOWN

1 Hand lens. (2 wds.)
2 The state of matter of ice.
3 To shove.
4 These hold plants in place. (roots leave)
6 The planets _____ the sun.
9 The quantity of something.
10 To discover a new tool or a new way to do something.
12 Measure how tall something is.
13 We look at a _____ to make a prediction about the future. (rhymes with Saturn)
15 The outside layer of a tree. (2 wds.)
17 When a plant species no longer exists.
18 What property could be used to compare pine needles and a maple leaf?
19 Something all living things need.
22 The green food making part of almost every plant. (rhymes with beef)

1. Which investigative question could be answered using one marble and an inclined plane?

O A. How does the release height on an inclined plane affect how far a marble will roll?
O B. How does the release height on an inclined plane affect how many marbles will roll?
O C. How does the release height on an inclined plane affect how well a marble will roll?

2. What kind of energy does a marble have just **before** the marble is released on an inclined plane?

O A. Kinetic energy

O B. Potential energy

O C. Chemical energy

3. Plan a new investigation using the following question: "At what height on the sloping track would you release a marble so that the marble would travel a distance of 3 meters from the **end** of the track?"

- Write your prediction.
- Write the steps for your procedure.

Use words, labeled pictures, and/or labeled diagrams in your answer.

Investigative Question: *At what height on the sloping track would you release a marble so that the marble would travel a distance of 3 meters from the* **end** *of the track?*

Prediction:

Materials: *marble, ramp, meter stick*

Procedure:

Lesson 7: "What would you attempt to do if you knew you could not fail?"
Dr. Robert Schuller, Minister

/ 25 = %

ACROSS

1 A liquid that carries oxygen to the cells in your body.
3 Organisms that eat other organisms. (consumer producer)
5 An animal's organ that pumps blood.
7 To change water from a liquid to a solid when cooled.
9 A person who studies science.
10 A place where condensed water vapor group together in the sky. (c____)
13 When something repeats over and over again.
15 Land that raises to great heights.
18 Snow, rain and hail are examples of _____. (condensation precipitation)
20 A simple machine that has a fulcrum.
21 The organ that covers the body of an animal.
22 Earth is a medium sized _____.
23 The Sun and everything on the Earth work together to form a weather _____. (s____)

DOWN

2 A measurement across the middle of a circle.
4 You use this to represent objects that are too big to explore. (m____)
6 Over 70% of earth is covered by this.
8 Sound bouncing back to the source.
11 1 pound = 16 _____.
12 Anything that can change in an investigation. (procedure variable)
13 To recycle plant material into nutrient rich soil.
14 When something takes place. (rhymes with cement)
15 A unit of length in the metric system.
16 A (spider, rabbit) is an example of a carnivore.
17 Precipitation that is a liquid.
19 To discover a new tool. (i____)

Parent/Guardian Signature:

For puzzle time, return on:

1. What research would **not** be useful in deciding which new fish species to add to a tank?

O A. The life cycle of the plant species.

O B. The predator/prey relationship of fish.

O C. The number of organisms the tank can hold.

2. What role is served by the bacteria that live between the rocks of this system?

O A. Producer

O B. Consumer

O C. Decomposer

3. What are two tools that students would use to take care of their new fish tank?

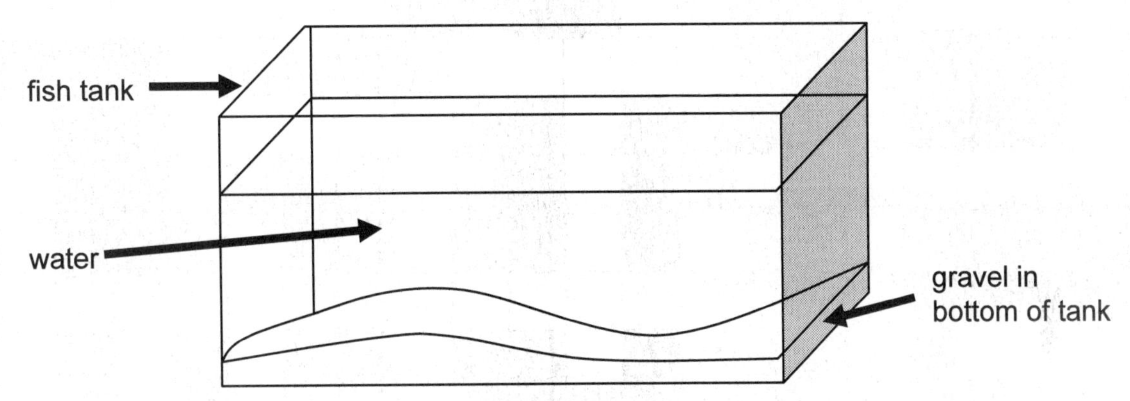

- Identify **two** tools to check the health of the tank system.
- Explain what these tools would measure.

Use words, labeled pictures, and/or labeled diagrams in your answer.

Tool:

What would this tool measure?

Tool:

What would this tool measure?

Lesson 8: "One of the secrets of life is to make stepping stones out of stumbling blocks."
Jack Penn, Author

/ 25 = %

ACROSS

1 The items used in an experiment. (m____)
4 Soil provides plants with mineral _____.
6 A living thing that can produce its own food through photosynthesis. (consumer producer)
8 Evidence of past animal life.
11 To use a substance again.
13 Kilograms are a measurement of _____.
14 The human organ used for chewing and breathing
15 Inputs, _____, and transfers of matter.
16 Animals that feed off of leftover food sources.
19 A place where animals live.
22 The process of protecting the Earth's resources. (rhymes with celebration)
23 Both a plant system and a human system need _____ to survive.
24 Movement of air.

DOWN

2 When offspring receive traits from their parents.
3 A plant part that grows on a stem.
5 This is passed onto other creatures when they eat plants.
6 Steps to complete an investigation. (rhymes with fan)
7 A place with little rain and plants.
9 The digestive _____ breaks down food in the body.
10 _____ power comes from the sun.
12 To describe why something happen. (e____)
17 These use carbon dioxide from the air and give off oxygen. (plants cougar lizard)
18 The reason for a flower is to produce _____. (s____)
20 The part of a plant that carries water and minerals from the roots.
21 Solid precipitation that falls from the sky. (snow rain)

1. While adding salt to water, the salt mixes with the water and is no longer visible. What has happened to the salt?

O A. A solution has been created.

O B. The salt has changed to its liquid form.

O C. The water has hidden the salt crystals.

2. No matter how much the water is stirred, why is salt seen at the bottom of a jar of salt water?

O A. The water has become a solid.

O B. The water has become saturated.

O C. The water has become condensed.

3. How is a lake's ecosystem affected by adding salt to the lake?

- Identify **two** possible effects on the lake system.
- Explain how plant and animal species could be affected.

Use words, labeled pictures, and/or labeled diagrams in your answer.

Effect on the lake:
How could this affect the plants?
Another effect on the lake:
How could this affect the animals?

Lesson 9: "The manner in which it is given is worth more than the gift."
Pierre Corneille, Playwright

/ 25 = %

ACROSS

3 This part of the plant contains the seeds. (2 wds.)
6 The ability to create offspring. (rhymes with introduce)
7 A living thing that feeds on other living things in a food chain. (consumer producer)
11 A series of events that happen over and over again.
14 When a liquid gets cold enough to become a solid it ______.
15 ______ are found in soil. (2 wds.)
18 The Earth casts its shadow on the moon during this event. (2 wds.)
21 Item that can be seen, touched, smelled, tasted and/or heard. (o____)
23 Vibrations in air or water.
24 The habitat in which an organism lives.
25 To show something. (rhymes with appreciate)

DOWN

1 A synonym for "procedures".
2 Africa, North America, and Asia are examples of ______.
4 All the living things in an area.
5 A metric measurement of mass.
8 Plants take in this gas. (2 wds.)
9 Plants produce this gas.
10 ______ and effect.
12 Water in its gas state. (2 wds.)
13 Measured in Celsius and Fahrenheit.
16 The act of sowing seeds in a garden. (used as a verb)
17 We feel the Sun's light energy in the form of ______.
19 A squeezing force.
20 The red liquid in human veins and arteries.
22 A measurement of distance.

Parent/Guardian Signature:

For puzzle time, return on:

1. Which of the following is a property of a chemical's pH?

O A. Acidity

O B. Turbidity

O C. Temperature

2. Which of the following is a physical property of a mineral that can be seen?

O A. Saltiness

O B. Hardness

O C. Luster

3. Explain the difference between a **chemical** change and a **physical** change.

- Give one example of a **chemical** change.
- Give one example of a **physical** change.

Use words, labeled pictures, and/or labeled diagrams in your answer.

Difference between a **chemical** change and a **physical** change:
Example of a **chemical** change:
Example of a **physical** change:

/ 25 = %

ACROSS

3 A squeezing force.
7 The general name of a producer.
9 Visible traits.
10 Animals that feed off of leftover food sources.
12 Metric measurement of mass. (k____)
17 Africa, North America, and Asia are examples of _____.
18 The ability to create offspring.
19 Anything that can change in an investigation.
20 To shove.
23 Vibrations in air or water. (s____)
24 A living thing that feeds on other living things in a food chain.
25 A liquid that carries oxygen to the cells in your body.

DOWN

1 The source of energy for the water cycle.
2 Measured in Celsius and Fahrenheit.
4 The state of matter of ice.
5 The cardiovascular _____ is responsible to bring in oxygen to the body.
6 Snow, rain and hail are examples of _____.
8 A simple machine that has a fulcrum.
11 Sound bouncing back to the source.
13 Soil provides plants with mineral _____.
14 This is passed onto other creatures when they eat plants.
15 A plant part that grows on a stem.
16 A living thing that can make its own food through photosynthesis. (consumer producer)
21 Measure how tall something is in inches, feet, and yards.
22 A place where condensed water vapor group together in the sky. (rhymes with proud)

1. What is the biggest danger to the ecosystem by releasing some classroom fish into a nearby lake?

O A. The fish could become a food source for local animals.
O B. The fish could eat the same food that the local fish eat, causing a shortage of food.
O C. The fish could adapt to their new environment over time.

2. Which category best describes the role of the trout in a river system?

O A. Producer

O B. Consumer

O C. Scavenger

3. How does removing trees from the river bank affect the ability of salmon to survive in a river system?

- Identify **two** effects on the river system.
- Explain how the salmon species could be affected.

Use words, labeled pictures, and/or labeled diagrams in your answer.

Effect:
How does this affect the salmon?
Another Effect:
How does this affect the salmon?

Lesson 11: "Light tomorrow with today!"
Elizabeth Barrett Browning, Author

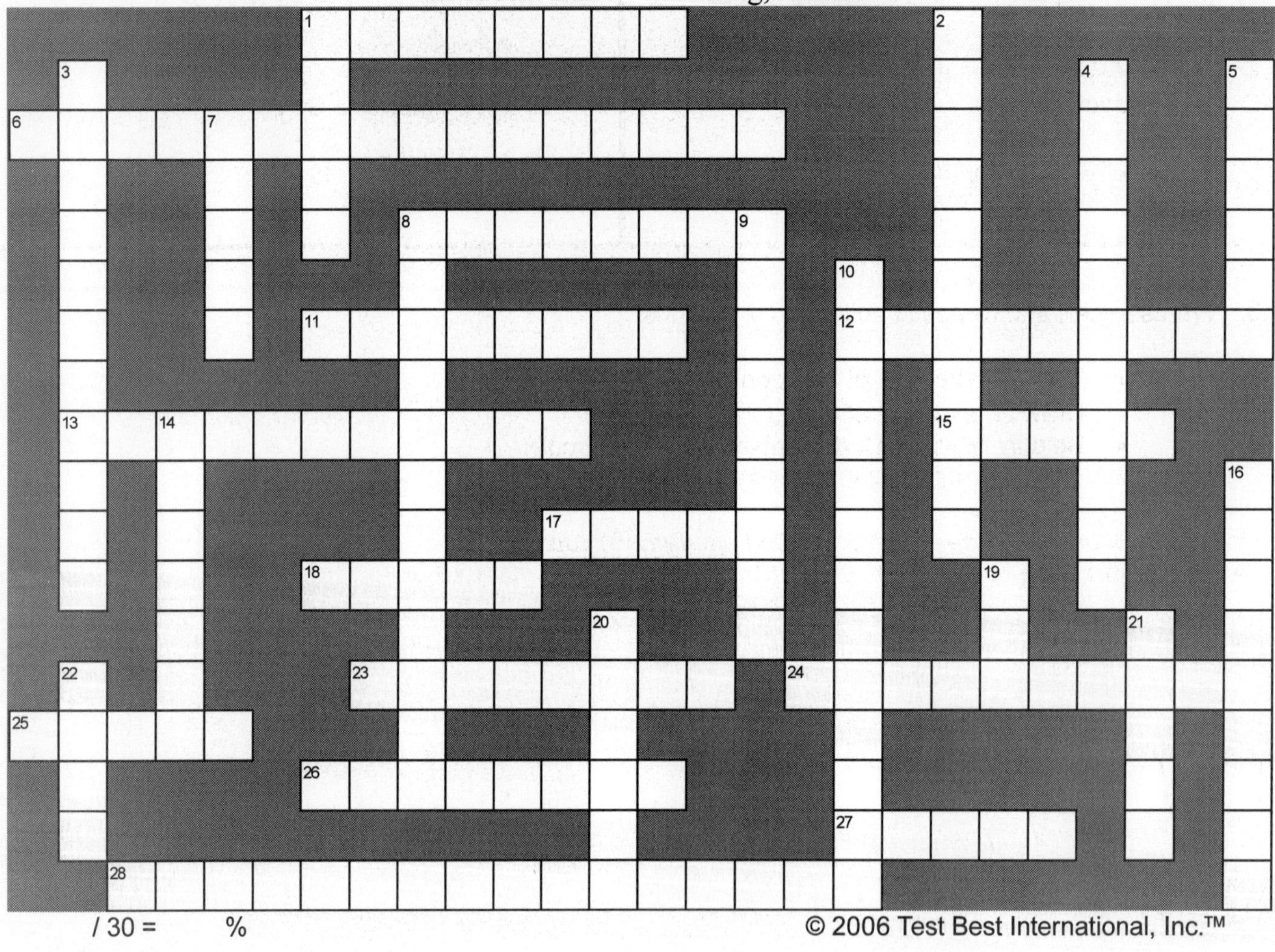

/ 30 = %

ACROSS

1 A landform that rises above the surrounding area.
6 This sound is created by fast vibrations. (3 wds.)
8 To save energy. (rhymes with deserve)
11 The study of how traits are passed between generations.
12 Things that are not made of cells.
13 The tool that measures the change in temperature.
15 The part of the plant that absorbs water and mineral nutrients from the soil.
17 A form of energy that can be seen. (l____)
18 A low-lying area that contains water.
23 When solids are mixed into a liquid and you can't see the solid anymore.
24 The repeated evaporation, condensation and precipitation on Earth. (2 wds.) (w____ c____)
25 A unit of weight in the standard system.
26 A tool used to magnify objects. (2 wds.)
27 A measurement of weight
28 Comparing where one object is to another object. (2 wds.) (r____ p____)

DOWN

1 1 _____ = 100 centimeters.
2 Energy caused by vibrations in air or water. (2 wds.)
3 The _____ system moves blood throughout the body.
4 Watching weather and recording is to make an _____. (o____)
5 The act of being alive. (rhymes with giving)
7 Describing how high or how low a sound is.
8 When water vapor changes into a liquid state. (condensation perspiration)
9 The interaction of the organisms and the environment in a given area. (e____)
10 The scientific process of answering a question.
14 Movement of weathered earth materials.
16 One of the steps to an experiment or investigation.
19 The mixture of gasses that surround the Earth.
20 _____ supply herbivores with food.
21 An incline plane wrapped around a center pole. (rhymes with blue)
22 Animals need plants as a source of _____.

Parent/Guardian Signature:

For puzzle time, return on:

1. What are materials called that allow electricity to move through them easily?

O A. Insulators

O B. Conductors

O C. Circuit

2. Sarah-Jo put some bread into a toaster. In order for the toaster to cook the bread, electrical energy changes into what form of energy?

O A. Chemical energy

O B. Sound energy

O C. Heat energy

3. We use open and closed circuits in everyday life.

- Draw an example of an open circuit.
- Draw an example of a closed circuit.
- Be sure to **label all the parts** of the two circuits.

Use words, labeled pictures, and/or labeled diagrams in your answer.

A labeled drawing of an OPEN CIRCUIT:

A labeled drawing of a CLOSED CIRCUIT:

Lesson 12: "The things that count most in life are usually the things that cannot be counted."
Bernard Meltzer, Attorney

/ 30 = %

ACROSS

2 Tool that helps you do work and transfers energy.
4 To dissolve the maximum amount of a substance into a solution. (rhymes with celebration)
8 Freezing, thawing, and then freezing again. (r____)
11 Evaluating the results. (rhymes with reviewed)
14 When the water overflows the banks of a river or stream.
20 A flowing stream of fresh water.
21 A thermometer measures the change in _____.
23 Solid, liquid and gas are examples of different _____. (3 wds.) (s____ o_ m____)
26 Where mineral nutrients are found
27 A moving object has this type of energy. (3 wds.) (e____ o_ m____)
28 Smoke is one type of air _____. (rhymes with solution)
29 A landform with water on three sides.
30 A gas found in our air.

DOWN

1 The up and down movement of water. (rhymes with slaves)
3 As far as we know the only planet in our solar system that contains life.
5 One part of the digestive system.
6 Liquid that falls from the clouds.
7 One type of insect that pollinate flowering plants.
9 Ramp.
10 Energy that is transferred from one organism to another is this process. (2 wds.) (rhymes with explain)
12 The flash of light during a rainstorm.
13 _____ is a property on how an object feels. (t____)
15 The Earth _____ the sun.
16 Total seconds that an object rolls. (3 wds.) (a____ o_ t____)
17 Evaporation may _____ clouds to form.
18 A _____ is an example of an omnivore.
19 A metric unit to measure mass.
22 Describing how low or high a sound is.
24 A picture or an idea that represents an object in a different scale. (m____)
25 A large standard measurement of distance.

1. Which investigative question could be answered using earthworms, dry soil, and moist soil?

O A. What temperature of soil do earthworms prefer?
O B. How much moisture do earthworms prefer?
O C. Do earthworms prefer sandy soil or rocky soil?

2. How are earthworms helpful to the environment?

O A. Earthworms provide moisture to the soil.

O B. Earthworms provide nutrients to the soil.

O C. Earthworms provide warmth to the soil.

3. Using only the materials below, plan a new scientific investigation involving earthworms.

- Write your prediction.
- Write steps for your procedure.

Use words, labeled pictures, and/or labeled diagrams in your answer.

Investigative Question: *How does the amount of light affect the number of earthworms visible?*

Prediction:

Materials: *earthworms, flashlight, box, lid for box, timer*

Procedure:

Lesson 13: "The times change and we change with them."

John Owen, Poet

1 2 3 4 5 6 7 8 9 10 11 12 13 14 15 16 17 18 19 20 21 22 23 24 25 26 27 28 29

/ 30 = %

ACROSS

3 These are caused by the tilt of the Earth's axis. (rhymes with reasons)

5 A giant ocean wave caused by an earthquake.

6 Visible light that contains all of the colors of the rainbow. (2 wds.) (w____ l____)

7 The items used in an investigation.

11 To define or demonstrate a concept. (rhymes with freight train)

13 The process of creating new generations. (rhymes with tax deduction)

18 To find an answer. (s____)

19 Sounds created by low energy vibrations (examples: tapping objects, whisper). (2 wds.) (s____ s____)

20 Outcome or output.

21 A _____ is an animal that preys on other animals.

23 A type of tree that has green leaves or needles all year.

24 To re-use a substance.

25 A tiny bit of a substance or a molecule. (p____)

26 Water in its gas form. (2 wds.) (rhymes with wrapping paper)

27 This part of the human body system that is used to receive energy.

28 What is an organized way to do something? (rhymes with jarn)

29 Tigers, eagles, and spiders are examples of _____.

DOWN

1 The place that a plant lives. (h____)

2 _____ can carry seeds away from the plant on their fur.

3 Referring to the length of an object.

4 The source of energy for all plants on the Earth.

8 To add energy into a system. (input output)

9 Energy that humans can see. (2 wds.) (l____ e____)

10 What is a thick growth of trees?

12 A _____ feeds on dead plants and animals.

14 A simple machine that has a wheel with a grove and a rope to pull.

15 The process of protecting natural resources. (c____)

16 The study of the Earth's climate. (rhymes with roman mythology)

17 Bees are an example of plant _____. (decomposers pollinators)

22 A unit for measuring a push or a pull.

1. Why does the moon appear to shine in the night sky?

O A. The moon makes light like the sun.

O B. The moon reflects light from the sun.

O C. The moon absorbs light from the sun.

2. What causes the different shapes or phases of the moon we see in the night sky every 28 days?

O A. As the moon orbits the earth we only see a section of the moon that is shining.

O B. As the moon orbits the earth the moon changes its shape.

O C. As the moon orbits the earth part of it is hidden behind the earth.

3. The moon is part of the solar system.

- Draw and label the phases of the moon.
- Be sure to include these words in your labeled diagrams where appropriate:
  - ✓ Full moon
  - ✓ First quarter moon
  - ✓ Third quarter moon
  - ✓ New moon

Use words, labeled pictures, and/or labeled diagrams in your answer.

Draw and label the phases of the moon:

Lesson 14: "Let no one ever come to you without leaving better and happier."
Mother Theresa, Missionary

/ 30 = %

ACROSS

1 A person who studies systems.
3 Sounds created by high energy vibrations (examples: banging objects, yelling). (2 wds.) (l____ s____)
6 Condensation, evaporation, and _____ are parts of the water cycle.
9 This takes blood from the human heart to the other body parts.
10 A place to record data from an investigation. (rhymes with fable)
11 Materials that settle out of liquids over time. (s_____)
12 Controlling variables during an investigation makes the investigation a ______. (2 wds.)
14 Measured in liters. (v_____)
15 A small metric measurement of volume. (centimeter milliliter)
17 This property attracts butterflies to some flowering plants. (c_____)
20 ______ is carried by red blood cells in the human body.
21 Decaying plant materials.
23 Consistent.
24 The human body system that brings oxygen into the body. (rhymes with circulatory)
26 Tiny living creatures that are food for many sea animals.
27 Vertebrate animals that have hair and care for their young.
28 These chemicals protect plants from insects.
29 Something that takes up space and has mass. (rhymes with batter)

DOWN

2 To clear-up a misconception or describe an outcome. (rhymes with education)
3 Measured in centimeters, meters, and kilometers. (l_____)
4 Organisms that eat only plants are called ______.
5 Controls the human nervous system.
7 Animal cells that are joined together to form a structure.
8 Where do seeds develop in a plant?
13 Water trapped in the surface of the Earth. (2 wds.) (g_____ w_____)
16 Lots of plants grow of this type in rain forests.
18 Ice and wood are examples of this state of matter.
19 When a rolling car slows down, its speed _______. (increases decreases)
22 This is a single form of life. (plankton organism)
25 The movement of ______ can cause erosion. (wind seed)

1. Which of the following could show the flow of energy in a real food chain?

O A. bird→insect→plant→cat

O B. insect→plant→bird→cat

O C. plant→insect→bird→cat

2. When you dissect an owl pellet you can find rodent bones and bird bones. What does this tell you about the owl?

O A. The owl is a producer.

O B. The owl is a carnivore.

O C. The owl is prey to other animals.

3. A food chain shows the food energy path from one organism to another organism.

- Using only the organisms below, create a food chain.

  ✓ insect
  ✓ plant
  ✓ deer
  ✓ cougar

- Label each organism in your food chain as a consumer, producer, or decomposer.

Use words, labeled pictures, and/or labeled diagrams in your answer.

/ 30 = %

ACROSS

4 A flowing stream of fresh water.
5 A picture or an idea that represents an object. (m____)
6 Referring to the length of an object.
7 Decaying plant materials.
9 A form of energy that can be seen.
10 As far as we know the only planet in our solar system that contains life.
11 The mixture of gasses that surround the Earth.
13 A _____ is an example of an omnivore.
15 This is a single form of life.
18 Materials that settle out of liquids over time.
20 _____ is carried by red blood cells in human bodies.
21 The movement of _____ can cause erosion.
22 Smoke is one type of air _____.
23 The interaction of the organisms and the environment in a given area. (e____)
24 This takes blood from the human heart to the other body parts.
25 The place that a plant lives.

DOWN

1 Controls the human nervous system.
2 The part of the plant that absorbs water and mineral nutrients from the soil.
3 Describing how low or high a sound is.
4 The human body system that brings oxygen into the body.
6 The source of energy for all plants on the Earth.
8 _____ supply herbivores with food.
12 Water in a gas form. (2 wds.) (w____ v____)
13 A tool used to magnify objects. (2 wds.)
14 _____ can carry seeds away from the plant on their fur.
16 These are caused by the tilt of the Earth's axis.
17 What is an organized way to do something?
18 An incline plane wrapped around a center pole.
19 A landform that rises above the surrounding area.
21 The up and down movement of water.

1. Which investigative question could be answered using **only** a golf ball and a meter stick?

O A. How long does a golf ball bounce when dropped from a height of 40 cm?
O B. How many times does a golf ball bounce when dropped from a height of 40 cm?
O C. How fast does a golf ball bounce when dropped from a height of 40 cm?

2. What happens when you drop a golf ball and a basketball at the same height at the same time?

O A. The basketball will hit the floor before the golf ball does.
O B. The golf ball will hit the floor before the basket ball does.
O C. The basketball and golf ball will hit the floor at about the same time.

3. Using only the materials below, plan a new scientific investigation using a frozen golf ball, a room temperature golf ball, and a meter stick.

- Write your prediction.
- Write the steps for your procedure.

Use words, labeled pictures, and/or labeled diagrams in your answer.

Investigative Question: *How does temperature affect the height that a golf ball bounces from 3 meters?*

Prediction:

Materials: *frozen golf ball, room temperature golf ball, meter stick*

Procedure:

Lesson 16: "Advice: It's more fun to give than to receive."
Malcolm Forbes, Business Leader

/ 30 = %

ACROSS

7 Sound that is bounced back.
9 When an animal species no longer exists.
10 This is a tool to dig a hole.
12 Going one way, then going another way. You are changing _____.
15 The tool used to measure force. (2 wds.) (s____ s____)
18 Another word for material. (s____)
19 A machine that does work.
21 _____ is a property of an object measured in kg.
23 When water travels from one area to the ocean.
26 Prairies can be called _____. (g____)
27 This equals 16 ounces.
28 A tool that separates the colors in white light.
29 Something that is repeating.
30 You use this to represent objects that are too small to be seen. (m____)

DOWN

1 To show or tell using many details.
2 Juice, soda and tea are examples of this state of matter.
3 The sweet smelling part of a flower that attracts pollinators.
4 Mice, insects, and rabbits eat plants, snakes eat mice, hawks eat rabbits, snakes and mice is an example of a
5 This carries blood to the human heart.
6 A plant growing from a seed to produce more seeds. (2 wds.) (rhymes with recycle)
8 Life, water, and rock are different types of _____. (c____)
11 Bears, humans, crows and cockroaches are _____.
13 This is an item that has mass and takes up space.
14 Rocks formed from lava or magma. (2 wds.) (i____ r____)
16 The Earth is the third closest _____ to the sun.
17 The air surrounding the Earth.
20 The length between two points.
22 The form of energy that is measured with thermometer. (2 wds.) (h____ h____)
24 A _____ is the opposite of a pull.
25 The point where a lever arm pivots.

1. Fossils are found in what type of rock?

O A. Igneous rock

O B. Sedimentary rock

O C. Metamorphic rock

2. What tool could help you find a fossil?

O A. Thermometer

O B. Magnifying glass

O C. Balance scale

3. Explain how a fossil of a **leaf** and **sea shell** can be found in a dry and bare desert.

- Explain how a fossil of a leaf could be formed in this area.
- Explain how a fossil of a sea shell could be formed in this area.

Use words, labeled pictures and/or labeled diagrams in your answer.

Leaf fossil:

Sea shell fossil:

Lesson 17: "Well done is better than well said."

Benjamin Franklin, Statesman

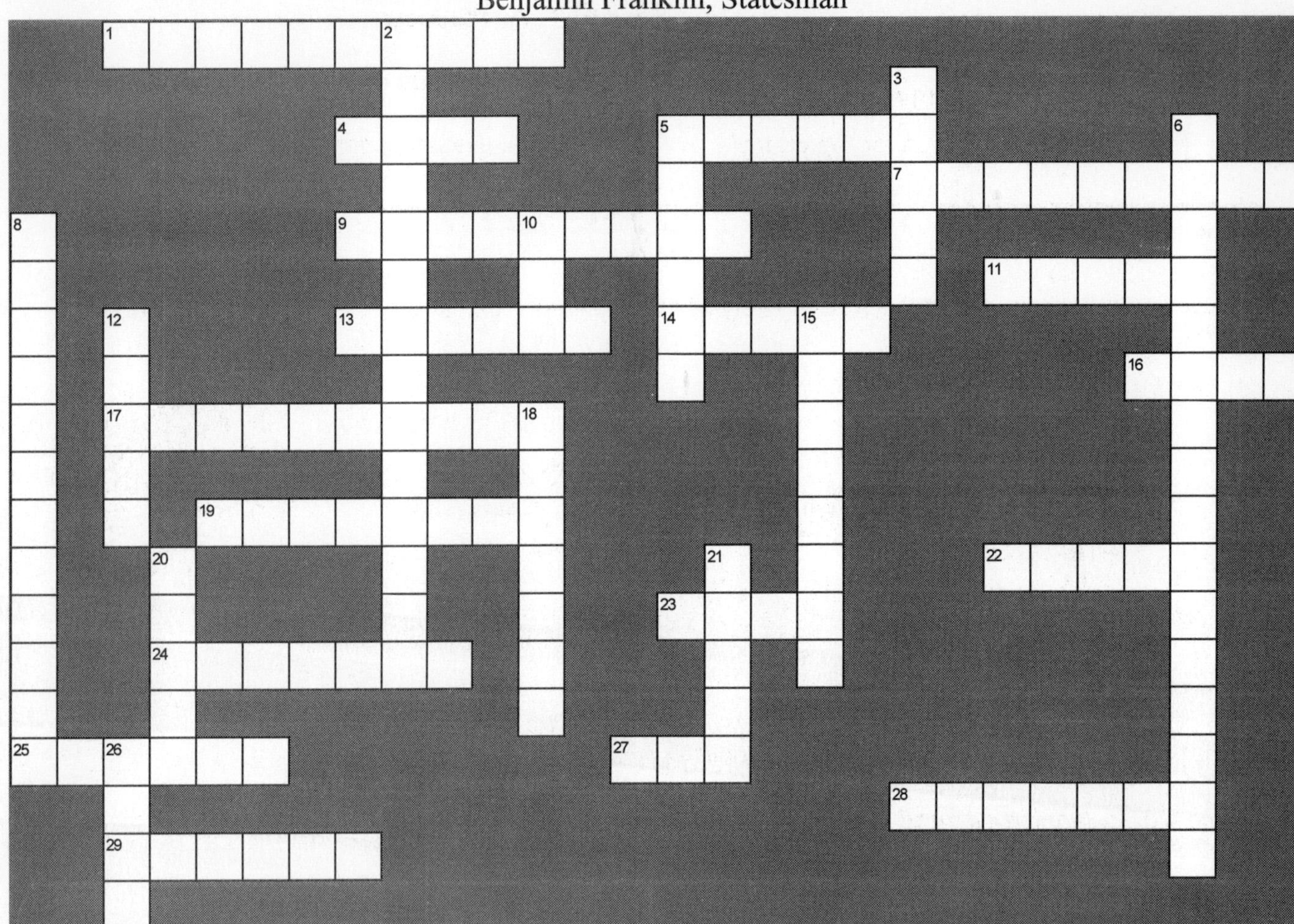

/ 30 = %

ACROSS

1 You can't hear sound in ______ because there is no matter for it to travel through. (2 wds.) (rhymes with meeting place)

4 A small body of still water.

5 The human tissue that moves bones.

7 The human body system that changes food into chemical energy.

9 Things that never grow and reproduce.

11 ______ and effect.

13 A ______ is made of parts that function as a whole. (s____)

14 A form of energy produced by the sun.

16 ______ forms as a result of condensation in the clouds. (rhymes with train)

17 Color, size, shape, and texture are all ______ of objects.

19 To change water vapor into liquid water. (c____)

22 A crack in the Earth's crust.

23 A natural satellite that rotates and revolves around the Earth.

24 An object in space casting a shadow on another object in space. (e____)

25 All living things on Earth depend on ______ for survival.

27 One of the three states of matter that takes both the shape and volume of its container.

28 The stem and roots provide ______ for the plant. (rhymes with report)

29 Measured in inches, feet, and yards. (l____)

DOWN

2 This process occurs in the leaf of a plant to supply energy.

3 Two incline planes back to back.

5 One of the layers of the Earth.

6 Plants need this from the soil to grow. (2 wds.)

8 Comparison or interaction between objects or events. (r____)

10 Water in its solid state.

12 Another word for gas. (v____)

15 A tool used to see things up close.

18 An object that is unique, unusual or distinctive. (s____)

20 A large body of salt water.

21 ______ absorb water and mineral nutrients from the soil.

26 Shaft that a wheel turns on.

1. Where does the energy flow chart begin in the diagram below?

O A. The soil provides nutrients for the carrot.

O B. The sun provides energy for the carrot.

O C. The carrot provides food for the bunny.

2. What is the role of the carrot in this system?

O A. Producer

O B. Consumer

O C. Decomposer

3. How would the removal of the wolves likely affect the other organisms in the energy flow chart?

- Identify **two** possible effects of removing the wolves from this system.
- Explain why these changes would likely happen.

Use words, labeled pictures, and/or labeled diagrams in your answer.

Effect:
Explain why this change would happen:
Another effect:
Explain why this change would happen:

Lesson 18: "We make ourselves rich by making our wants few."
Henry David Thoreau, Philosopher

| 1 | 2 | 3 | 4 | 5 | 6 | 7 | 8 | 9 | 10 | 11 | 12 | 13 | 14 | 15 | 16 | 17 | 18 | 19 | 20 | 21 | 22 | 23 | 24 | 25 | 26 | 27 | 28 | 29 | 30 |

/ 30 = %

ACROSS

5 A very large weather event that forms over warm water. (h_____)
8 When something happens. (e_____)
9 One part of the digestive system.
10 The name of organisms that make their own food.
13 ______ energy comes from electrons moving.
14 A sprout is like this stage in a human's life. (rhymes with maybe)
15 The smallest unit of living creatures.
16 Nutrient minerals for plants are like ______ for the human system. (v_____)
17 The process of passing traits to future generations.
19 Photosynthesis occurs in the plant's ______.
20 A collection of liquid water (lakes creek brook)
23 A natural crystal material.
24 Tiny ice crystals that fall from clouds. (s_____)
25 Lots of these are on our earth. (rhymes with sock)
27 A change of direction of a sound wave. (rhymes with protection)
28 The scientific guess of the outcome of an investigation.
29 A ______ is an example of a producer.
30 A measurement of mass.

DOWN

1 A colorless liquid that carries nutrients.
2 The regular movement of Earth's water caused by the moon.
3 Plant stems can move in the wind because they are ______. (f_____)
4 The energy or work produced by a human. (output inputs)
6 The process by which rocks are formed and changed. (2 wds.)
7 To change the measurement from meters to yards. (c_____)
11 The energy that comes from a chemical change. (2 wds.)
12 One of the four simple properties of an object. (s_____)
18 A landform surrounded by water on all sides.
21 Water with another substance added.
22 A tiny bit of matter.
26 A large body of salt water within an ocean.

Parent/Guardian Signature: For puzzle time, return on:

1. Which of the following is true about magnets?

O A. Magnets with the same poles will attract each other.

O B. Magnets with opposite poles will push each other away.

O C. Magnets with the same poles will push each other away.

2. How could the magnet below be used to sort materials?

O A. Separate aluminum foil from gum

O B. Separate paperclips and crayons

O C. Separate insects from plants

3. Besides sorting materials, identify a function of magnets in human lives and describe how the magnet has helped people.

S N

Use words, labeled pictures, and/or labeled diagrams in your answer.

Function of Magnet:

How has this made people's lives better?

Lesson 19: "It is never too late to become what you might have been."
George Eliot, Author

/ 30 = %

ACROSS

1 When organisms continue to live through hard times.
2 Tiny drops of water in the air.
5 This is a visual aide in a diagram.
9 All of the light energy you can see and the light energy you can't see. (2 wds.) (l____ s____)
10 The part of the flower where seeds begin to develop. (o____)
12 Tiny living creatures that live in salt water.
13 Single celled organisms that can only be seen through magnification. (b____)
14 Melted rock above the Earth's surface.
15 A general term to explain a structure inside an organism that has a specific function.
16 Something that needs to be solved.
19 Light that bounces off of an object. (2 wds.) (r____ l____)
21 An illness that can spread between plants and cause harm.
23 When liquid water changes into water vapor.
26 A lens that makes small objects look larger. (2 wds.)
28 A pair of breathing organs.
29 When pollen is transferred from the pistil to the stamen of the flower. (rhymes with communication)

DOWN

1 To classify or organize. (rhymes with fort)
3 Hyenas and vultures are ______.
4 A ______ is the opposite of a push.
6 Traits that are recognizable.
7 When seeds sprout we call this ______.
8 A ______ is anything that can change in an experiment or investigation.
11 Cause and ______.
17 To show is to ______. (d____)
18 What happens to water that reaches 0° Celsius?
20 To change. (t____)
22 The cover and protection for an animal. (s____)
24 A part of a flower.
25 Used to measure length. (i____)
27 Solid form of water.

1. What are the smallest units of life called?

O A. Plankton

O B. Oxygen

O C. Cells

2. How are plant cells and animal cells the same?

O A. They both have a nucleus.

O B. They both do photosynthesis.

O C. They both contain chloroplast.

3. A cell is a system made of parts that work together.

- Draw and label the parts of a plant cell.
- Draw and label the parts of an animal cell.
- Be sure to include these words in your labeled diagrams where appropriate:
  - ✓ cell wall
  - ✓ cell membrane
  - ✓ nucleus
  - ✓ cytoplasm
  - ✓ chloroplasts

Use words, labeled pictures, and/or labeled diagrams in your answer.

Labeled diagram of a plant cell:

Labeled diagram of an animal cell:

Name: Lesson 20: Assessment 4

/ 30 = %

ACROSS

3 The process by which rocks are formed and changed. (2 wds.)
9 A natural crystal material.
12 An illness that can spread between plants and cause harm.
13 Single celled organisms that can only be seen through magnification.
14 A large collection of water.
15 The sweet smelling part of a flower that attracts pollinators.
16 A collection of liquid water (lakes creek brook)
17 The stem and roots provide _____ for the plant.
19 The name of organisms that make their own food.
21 Things that never grow and reproduce.
26 Used to measure length.
27 Color, size, shape, and texture are all _____ of objects.
28 The air surrounding the Earth.

DOWN

1 Water in its solid state.
2 Mice, insects, and rabbits eat plants, snakes eat mice, hawks eat rabbits, snakes and mice is an example of a
4 The energy that comes from a chemical change. (2 wds.)
5 A form of energy produced by the sun.
6 A colorless liquid that carries nutrients.
7 The scientific guess of the outcome of an investigation.
8 _____ and effect.
10 Comparison or interaction between objects or events.
11 To classify or organize.
17 When organisms continue to live through hard times.
18 This equals 16 ounces.
19 Something that needs to be solved.
20 A _____ is an example of a producer.
22 This is an item that has mass and takes up space.
23 A pair of breathing organs.
24 A machine that does work.
25 A _____ is the opposite of a pull.

1. What is the purpose of the string in the balloon rocket system below?

O A. Control the flow of energy

O B. Control the speed of the rocket

O C. Control the direction of the flight

2. What force causes the rocket to move forward after being released?

O A. The act of moving the handle of the air pump back and forth.

O B. The air inside the balloon pushing against the outside air.

O C. The balloon is being pulled by the string.

---

3. How can the rocket be changed to make the rocket fly farther on the string?

- Identify **two** changes that could be made to the rocket system to improve flight distance.
- Explain how these changes would allow the rocket to fly farther.

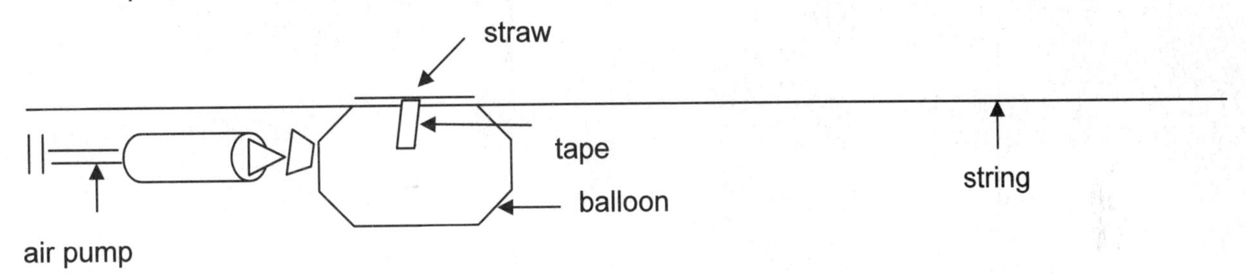

Change to rocket:
Effect of the change:
Change to rocket:
Effect of the change:

Lesson 21: "Before I speak, I have something important to say."
Groucho Marx, Comedian

/ 35 = %

ACROSS

4 The form of energy once an object is raised to a higher level. (2 wds.) (p____ e____)
6 If an animal is a male, that is the animals ____. (rhymes with blender)
8 A human uses their _____ to sense touch. (rhymes with thin)
9 Evidence of organisms from the past.
10 Precipitation comes from these objects in the sky.
12 A force is a push or a _____.
14 A small, flowing collection of water that may empty into a river.
16 An illness that can spread between organisms and cause harm. (rhymes with Chinese)
22 Animals that have hair and care for their young.
23 One type of reproductive cell in an organism.
24 This general term is measured using cups, quarts, and gallons. (v____)
26 Measuring an object vertically is that object's ____.
27 The part of the water cycle where rain falls.
28 Drops of liquid water falling from clouds.
29 The organ that allows animals to process information.
30 An organism that feeds on dead materials. (d____)
31 If you want more of something, you must _____ the mass (increase decrease).
32 The point at which a solution can no longer dissolve a substance.
34 ____ is a natural source of energy needed by plants for photosynthesis.
35 This process allows plants to make food using sunlight. (p____)

DOWN

1 A transparent and convex shaped lens. (2 wds.)
2 The 7 large land forms on Earth.
3 A measurement about the length of a forearm .
5 A general term for any substance that can flow freely. (l____)
7 Heart, lungs, skin, and brain are examples of an _____.
11 The moon changes in a predictable _____. (rhymes with Saturn)
13 Melted rock that comes out of a fault.
15 A general term for diagram showing how a process or a system works is called a ____. (rhymes with waddle)
17 A general term for sand, dirt and clay.
18 The process when organisms change over long periods of time. (e____)
19 A general term for anything that can change in an experiment.
20 The process of creating more organisms of the same kind.
21 A general term for these changes in the moon; crescent, quarter, full, and new.
25 A general term for rocks, aluminum, and glass for example. (n____)
33 The gases that surround the Earth.

1. How does the skin protect the apple?

O A. The skin protects the apple from drying out.

O B. The skin protects the apple from being eaten.

O C. The skin protects the apple from falling off the tree.

2. What variable is being measured in an investigation that involves a chopped apple, a peeled apple, and an unpeeled apple exposed to the air and being weighed every day for 5 days?

O A. The amount of water precipitation from each apple.

O B. The amount of water condensation from each apple.

O C. The amount of water evaporation from each apple.

3. The water cycle is an earth system.

- Draw and label the water cycle.
- Be sure to include these words in your labeled diagrams where appropriate:
  - ✓ Evaporation
  - ✓ Condensation
  - ✓ Precipitation
  - ✓ Runoff
  - ✓ Lake
  - ✓ Clouds
  - ✓ Mountain

Use words, labeled pictures, and/or labeled diagrams in your answer.

Labeled diagram of the water cycle:

Lesson 22: "A problem is a chance for you to do your best."

Duke Ellington, Musician

/ 35 = %

ACROSS

1 Animals that eat meat.
6 The study of the Earth. (2 wds.) (e____ s____)
7 Wind and flowing water often cause this.
8 A general term for the change in the amount of light, temperature and weather throughout the year. (rhymes with reasons)
11 A general term for a large body of salt water.
12 This process is how plants make their own food.
14 A small metric unit to measure weight.
15 Shape is just one of many different ____ of an object. (p____)
16 Plants need carbon dioxide, _____, and water for photosynthesis. (s____)
17 Measurements of weather over time.
18 The _____ cycle is a process in which rocks are formed and changed.
23 The type of energy that an organism gets from eating. (2 wds.) (f____ e____)
25 The breaking and wearing away of rock.
27 Apply this to plants to give mineral nutrients.
29 The largest organ of the human body.
30 The lever arm rests on the _____.
31 This type of cell can carry messages between an animal's brain and their muscles. (n____)
32 A solid, natural, Earth material with its own set of properties.
33 A form of energy from the sun.

DOWN

1 Organisms that get their energy by eating other organisms are _____. (consumers producers)
2 These carry blood to the heart of vertebrate animals.
3 This can be measured in ounces and pounds and is different then weight.
4 The general measurement term that describes seconds, days, and centuries.
5 The act of having life. (l____)
9 This can cause a habitat to be dirty, noisy, or unhealthy for an organism.
10 A newly designed or developed tool. (rhymes with detention)
12 Rain forms around a _____ of dust. (p____)
13 Solid, liquid, and gas are the three _____. (3 wds.)
19 A general term for anything that is produced by a plant; for example oxygen or fruit. (inputs output)
20 A cow and a sparrow are examples of this kind of animal. (carnivore herbivore)
21 Another term describing when an object takes in or soaks up.
22 A _____ can make its own food from sunlight.
24 Developing a plan to solve a problem. (d____)
26 A general term describing where a plant or organism lives.
28 Another word for push away.

Parent/Guardian Signature:

For puzzle time, return on:

1. Which of the following tools is an example of a lever system?

O A.

O B.

O C.

2. Which of the following is an example of using an inclined plane?

O A. Using a board to roll a stone up into the back of a pick up truck.

O B. Using a wheelbarrow to haul dirt.

O C. Using a hammer to drive nails into a board.

3. How are pulleys helpful to people?

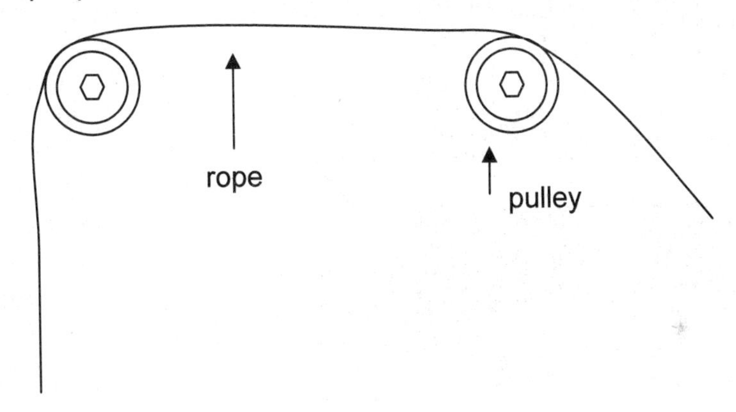

- Identify how a pulley is useful to people.
- Explain how the pulley system works.

Use words, labeled pictures, and/or labeled diagrams in your answer.

What is a pulley used for?
How does the pulley work?

Lesson 23: "Obstacles are things a person sees when he takes his eyes off his goal."
E. Joseph Cossman, Salesman

/ 35 = %

ACROSS

3 A general term for light energy you can see. (2 wds.)
6 When you mix sugar into water and the sugar it looks like it disappears.
7 Another word for a labeled picture.
11 A general term for warm-blooded vertebrate organisms that have fur or hair.
13 A general term for any living thing.
14 A word meaning to classify or put in order a set of objects or organisms. (o_____)
17 ______ is a property of a rock or mineral that is measured through a scratch test.
19 ______ are a part of a plant that absorbs mineral nutrients from the soil.
20 The term used to describe a lens changing the direction of light (bends light).
22 A general term used to describe organisms that are eaten for food by predators.
23 Unlike solids, ______ take on the shape of their container.
25 The food factory of a plant is a ______.
27 ______ are small organisms used for food by many sea animals.
29 This can be measured in ounces, pounds or grams and is different than mass.
30 The process of liquid forming on the outside of a cold glass of water. (rhymes with celebration)
34 Organisms that only eat animals are called ______.
35 If plants don't have water, they die because they need water for this process.

DOWN

1 The width of a sphere or ball.
2 A property of the surface of an object.
4 To change a liquid into a gas.
5 The general term used to describe reporting the data of an investigation in a concise way.
8 The general term for the gasses we breathe.
9 It is necessary for plant growth and is found in soil. (2 wds.)
10 Many plants grow from ______.
12 Questions, predictions, materials list, procedures, and data are parts of an ______.
15 The skeletal system is made up of these.
16 The type of energy made when objects vibrate.
18 The food chain shows ______ moving through the ecosystem. (e_____)
21 Evaporation, condensation, and precipitation are stages of the water ______.
24 This type of organism uses the process of photosynthesis to make its own food. (consumer producer)
26 Food chains that are connected and overlap in a ecosystem are called ______. (2 wds.)
28 The force that keeps the planets orbiting the sun.
31 A general term for the property of an object that can describe its volume, mass or weight. (s_____)
32 A general term for leaves, straw or wood chips which can be placed around a plant for protection. (m_____)
33 When a normally dry environment is covered with water.

Parent/Guardian Signature: For puzzle time, return on:

1. Which question could be answered using **only** ants and several food items?

O A. How fast do ants eat food?

O B. What types of food ants prefer?

O C. At what temperature do ants eat more food?

2. How can you tell an ant is an insect?

O A. Ants have 6 legs and two body parts.

O B. Ants have 6 legs and four body parts.

O C. Ants have 6 legs and three body parts.

3. Using only the materials below, plan a new scientific investigation involving ants.

- Write your prediction.
- Write steps for your procedure.

Use words, labeled pictures, and/or labeled diagrams in your answer.

Investigative question: *How does soil temperature affect how long it takes for ants to eat a 1 cm cube of cheese?*

Prediction:

Materials: *ants, warm soil, room temperature soil, cold soil, large box to hold soil, timer, cheese, ruler*

Procedure:

Lesson 24: "Nothing lasts forever, not even your troubles."
Arnold H. Glasgow, Psychologist

/ 35 = %

ACROSS

6 An illness that can spread between humans and cause harm.
7 The general term for the type of energy from the sun.
8 A general term for the results of a cause or producing an outcome.
10 Organisms that move and eat other organisms.
12 The general term for the different types of places plants and animals can live.
15 People are _____ beings.
20 A small unit of length, about the length of a thumb.
22 This part of the plant catches sunlight to make food for a plant.
24 The act of saving energy.
25 This plant part provides anchoring for the plants during windy weather.
26 The _____ system provides the body with oxygen.
29 An organism that breaks down dead material.
31 This is a part of the plant that can make seeds.
32 The process of watch something closely and record data.
33 _____ supply animals with oxygen. (p____)
34 To _____ is to re-use an object.
35 The digestive _____ is a process of breaking down food for energy.

DOWN

1 The sun's energy is a source of _____ . (l____)
2 This tool is used to see small things. (2 wds.)
3 The smallest unit of living systems.
4 Lions, sharks, and eagles are examples of _____.
5 The process of moving earth materials. (rhymes with explosion)
9 Plants get _____ and water from the soil. (2 wds.)
11 The human organ that holds and partially digests food.
13 To change the measurement from centimeters to inches.
14 This provides cover and protection for a organism.
16 We use ____ to represent systems that cannot be explored in a classroom.
17 A _____ is anything that can change in an investigation.
18 Another word for image or drawing.
19 The study of how traits are passed from parent to offspring.
21 Plants need _____, sunlight and water for photosynthesis. (2 wds.) (c____ d____)
23 Plants and animals are mostly made of this kind of liquid.
27 A general term meaning to arrange object or put objects in order.
28 The human system receives energy from _____.
30 When a species no longer exists that species is ____.

Parent/Guardian Signature:

For puzzle time, return on:

1. How do ladybugs help plants?

O A. Ladybugs eat aphids.

O B. Ladybugs crawl on leaves.

O C. Ladybugs lay eggs on plants.

2. How many body parts and legs does a ladybug have?

O A. Two body parts and 6 legs.

O B. Three body parts and 6 legs.

O C. Four body parts and 6 legs.

3. A ladybug is a living system made of specialized parts that work together.

- Label the parts of the ladybug system below.
- Select one part of the ladybug and describe its function.

Label the parts of the ladybug system below using the following terms: Eye, head, antennae, leg, elytra, pronotum, spot

Ladybug system part selected:
Function of the part selected:

Name: Lesson 25: Assessment 5

1 2 3 4 5 6 7 8 9 10 11 12 13 14 15 16 17 18 19 20 21 22 23 24 25 26 27 28 29 30 31 32 33 34

/ 35 = %

ACROSS

1 The part of the water cycle where rain falls.
4 Evidence of organisms from the past.
7 This plant part provides anchoring for the plants during windy weather.
10 _____ is a natural source of energy needed by plants for photosynthesis.
11 A word meaning to classify or put in order a set of objects or organisms.
12 To _____ is to re-use an object.
13 The lever arm rests on the _____.
14 Another term describing when an object takes in or soaks up.
16 The study of how traits are passed from parent to offspring.
18 A _____ is anything that can change in an investigation.
21 A general term for anything that is produced by a plant; for example oxygen or fruit.
22 The moon changes in a predictable _____.
24 Another word for a labeled picture.
25 The act of having life.
27 The act of saving energy.
29 The skeletal system is made up of these.
30 A general term describing where a plant or organism lives.
32 A general term used to describe organisms that are eaten for food by predators.
33 The process of creating more organisms of the same kind.
34 This tool is used to see small things. (2 wds.)

DOWN

1 This process is how plants make their own food.
2 The process of moving earth materials.
3 A cow and a sparrow are examples of this kind of animal.
5 The process of watch something closely and record data.
6 Heart, lungs, skin, and brain are examples of an _____.
8 A general term for sand, dirt and clay.
9 A general term for the property of an object that can describe its volume, mass or weight.
15 An organism that feeds on dead things.
17 _____ supply animals with oxygen.
19 The organ that allows animals to process information.
20 _____ are organisms used for food by many sea animals.
23 The food chain shows _____ moving through the ecosystem.
26 The type of energy made when objects vibrate.
28 The _____ cycle is a process in which rocks are formed and changed.
31 The food factory of a plant is a _____.

Parent/Guardian Review: Date:

1. Which investigative question could be answered using vinegar, baking soda, and red cabbage juice?

O A. How does red cabbage juice change to indicate an acid or a base?
O B. How does red cabbage juice change to indicate sweet or sour?
O C. How does red cabbage juice change to indicate a solid or a liquid?

2. What happens when you add an acid and a base together?

O A. A physical reaction occurs.

O B. A chemical reaction occurs.

O C. A liquid reaction occurs

3. Using only the materials below, plan a new scientific investigation adding lemon juice that is hot, cold, and room temperature to baking soda.

- Write your prediction.
- Number the steps to your procedure.

Use words, labeled pictures, and/or labeled diagrams in your answer.

Investigative Question: *How does the amount of baking soda added to lemon juice affect the temperature of the solution?*

Prediction:

Materials: *lemon juice, baking soda, timer, thermometer, jars, measuring spoons and cups, safety glasses*

Procedure:

Lesson 26: "If opportunity doesn't knock, build a door."
Milton Berle, Comedian

/ 35 = %

ACROSS

1 The general term used for warm-blooded vertebrate animals that have hair and care for their young.
6 A general term for all the observations or data that prove a conclusion.
7 Earth's most common liquid.
8 Earth receives most of its _____ from the sun.
10 Producers make their own ______.
12 The lungs are part of the ______ system.
14 Another word for explore, examine or question.
19 The type of organism that uses the sun to do photosynthesis.
20 Seeds from plants are sometimes spread by ______.
22 The star that supplies energy to Earth.
24 ______ is the property that describes the form of an object.
25 Anything that lives is an ______.
26 System of bones that support the human body.
28 The general term for receiving physical characteristics from an organism's parents. (i____)
30 There are nine ______ that orbit the sun.
31 To change water into a gas. (e____)
32 Future generations of plants can be called the plants' _____.
33 The process of having too much water in a habitat.

DOWN

2 Sound can only travel through ______ such as air, water, and solid objects. (rhymes with batter)
3 The term used to describe measuring the heaviness of an object.
4 The part of the plant where photosynthesis takes place.
5 An animal that is hunted and eaten by another animal.
9 The force between 2 objects.
11 A labeled picture showing how a system works.
13 The term used to describe large amounts of ice moving across land.
15 The tool used to measure weight. (2 wds.) (s_____ s_____)
16 An organism that receives its energy by digesting other organisms. (rhymes with good humor)
17 We use this to represent systems that are too small to be seen.
18 The colorful and fragrant part of the flower that attracts an organism.
19 A general term for the steps in an investigation.
21 Finding a total number of objects is the same as the ______ of objects. (a____)
23 A person who studies the natural or physical world.
24 The respiratory ______ is where air is taken into the body.
27 An animal's breathing organs.
29 The scientific unit for measuring "a push" or "a pull." (n_____)

1. In what part of the tomato plant system does reproduction happen?

O A. Seed

O B. Leaves

O C. Flowers

2. What makes a tomato plant a producer?

O A. The ability to grow leaves and fruit.

O B. The ability to make its own food.

O C. The ability to reproduce.

---

3. How would the tomato plant system be affected if it could no longer complete one stage of the life cycle?

Flower (not shown)

Fruit

Leaf

Stem

Roots (inside pot)

- Identify the stage of the life cycle that can no longer be done.
- Explain how this affects the ability of the tomato plant to live and grow.

Use words, labeled pictures, and/or labeled diagrams in your answer.

Incomplete life cycle stage:
Effects on the tomato plant's ability to live and grow:

Lesson 27: "To be upset over what you don't have is to waste what you do have."
Ken S. Keyes, Jr., Author

/ 35 = %

ACROSS

2 The sprouting of a plant from a seed.
7 A large metric measurement of distance.
11 The process of changing the measurement from liters to quarts.
13 A mountain created from an opening in the earth's surface.
16 Changes in organisms that occur over long periods of time. (rhymes with resolution)
18 The parts of the plant that takes in carbon dioxide from the air.
20 A general term used to describe units of measurement such as grams and kilograms.
22 The opposite of dieing.
23 Earth, Mars, Jupiter, and Venus are examples of different _____.
25 A scientific unit for measuring "a pull".
26 Ice crystals that fall in soft white flakes.
28 Bats and butterflies act as ______ for flowers.
31 The material that surrounds the roots of most plants.
32 A digging tool that is also a lever.
33 This is a small unit of weight in the metric system.
34 Another word for frozen water.
35 A gas exhaled by humans needed for plant survival. (2 wds.)

DOWN

1 The general term for giving instructions to get to one place to another.
3 Another word describing how organisms are connected to each other. (r____)
4 When an object takes in light, that light has been ______by the object.
5 A general term for the steps to answer a question. (p____)
6 A general term used to describe units of measurement such as seconds, minutes and hours.
8 To ______ is to use an object again.
9 The general term for everything in a habitat that surrounds an organism. (e____)
10 The general term to describe the moon moving around the earth.
12 An insect eats leaves, a bird eats the insect, then a hawk eats the bird is an example of a ______. (2 wds.)
14 A general term for small pieces of earth materials found in the bottom of rivers or lakes.
15 Trees that loose their leaves in the fall are called ______ trees.
17 A small, continuous flow of water.
19 To arrange objects from largest to smallest is an example of ____.
21 The study of the Earth's weather.
24 This plant part holds up the leaves and flowers.
27 The process that changes objects from a solid to a liquid by heating.
29 A gas that is important for survival of animal life.
30 A general term that describes a rapidly moving body of fresh water.

Parent/Guardian Signature: For puzzle time, return on:

1. What tool would you use to measure the distance a paper airplane flies when thrown?

O A. Meter stick
O B. Spring scale
O C. Thermometer

2. What is one force that is acting on an airplane while in flight?

O A. Magnetic force
O B. Frictional force
O C. Electrical force

3. Justin and Chelsea predicted that a paper airplane with larger wings would fly a farther distance than an airplane with smaller wings. They decided to do an investigation and collected the data below.

Airplane Wing Size	**Distance the Paper Airplane traveled in meters (m)**			
	Trial 1	**Trial 2**	**Trial 3**	**Average**
Small	8	10	9	9
Medium	7	5	6	6
Large	5	5	5	5

Write a conclusion for Justin and Chelsea's investigation. Be sure to:

- Tell whether Justin and Chelsea's prediction was correct or not.
- Include their results of high and low data averages in your explanation.
- Compare the results of high and low data averages.

Tell whether the prediction is correct or not:

Include high and low data averages:

Compare the high and low data averages:

Lesson 28: "You have to have a dream so you can get up in the morning."
Billy Wilder, Actor

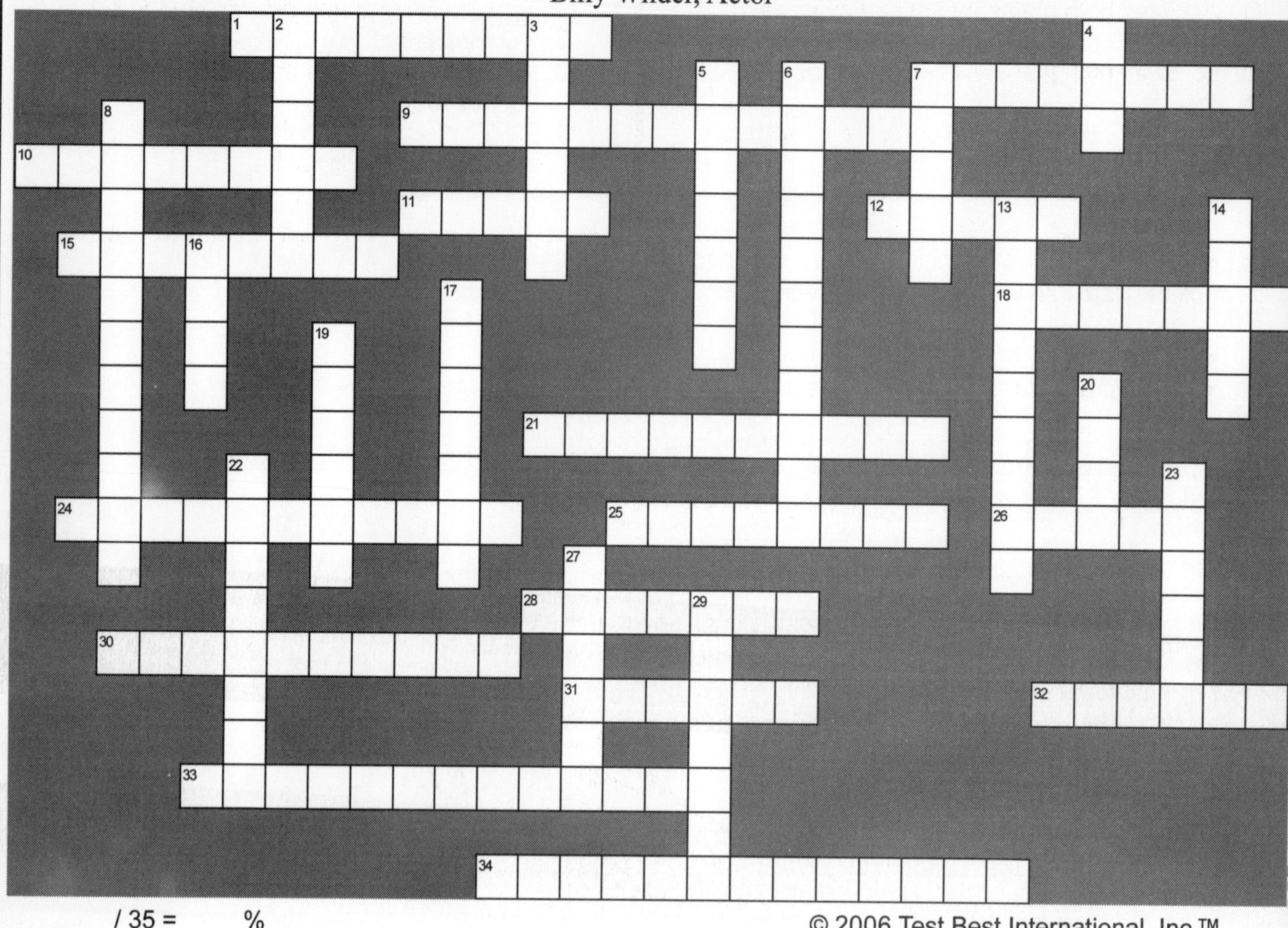

/ 35 = %

ACROSS

1 Humans are _____ because they eat plants and animals.
7 Humans are a (producer consumer).
9 When an object moves it has _____. (2 wds.) (k_____ e_____)
10 The luster is a physical _____ of an object.
11 Humans' breathing organs.
12 The liquid that carries nutrients to the different parts of the body.
15 The property of a mineral to resist scratching. (h_____)
18 The force that causes rain to fall to Earth.
21 When the Earth's crust moves, it causes an _____.
24 The system that provides the body with oxygen.
25 The _____ protects the inner organs of an animal.
26 We use this to represent events that cannot be easily manipulated.
28 A type of weather that happens again and again is called a weather _____.
30 Logical and systematic thinking can be called _____ reasoning. (s_____)
31 A scientific unit for measuring force or work.
32 Evidence of past organism life.
33 Rain, snow, sleet, hail are examples of _____.
34 Another word for an experiment that tests a prediction.

DOWN

2 The melted rock layer below the Earth's crust.
3 The organisms needs _____ so it can do work.
4 Producers get their energy from the _____.
5 The general term for the conditions of the atmosphere in a local area.
6 The type of measurement used in weather and measured in degrees.
7 A repeating series of events or actions.
8 The Sun and all the objects that orbit the sun. (2 wds.)
13 A general term which insects, plants, and humans are example.
14 The _____ is how high or how low a sound tone is.
16 The information that needs to be recorded during an investigation.
17 The general term to stake or to keep plants upright. (rhymes with report)
19 The liquid inside a flower.
20 The energy source for organisms.
22 A measurement across the middle of a ball or sphere.
23 This type of organism can give animals shelter from weather. (p_____)
27 Metal that may attract other pieces of metal.
29 When all organism of a species die they are considered _____.

Parent/Guardian Signature:

For puzzle time, return on:

1. How are bees helpful to plants?

O A. Bees help plants by eating flower nectar.

O B. Bees help plants by making honey.

O C. Bees help plants by pollinating flowers.

2. How are flowering plants helpful to bees?

O A. Flowering plants provide nectar and pollen for bees.

O B. Flowering plants provide protection and warmth for bees.

O C. Flowering plants provide shelter and camouflage for bees.

3. A bee is a living system made of specialized parts that work together.

- Label the parts of the bee system below.
- Select one part of the bee system and describe its function.

Label the parts of the bee system below using the following terms: Compound eye, head, antennae, leg, thorax, abdomen, wing, proboscis

Bee system part selected:
Function of the part selected:

Lesson 29: "Even if you're on the right track, you'll get run over if you just sit there."
Will Rogers, Humorist

/ 35 = %

ACROSS

2 Anything with parts working together, inputs, and outputs. (s____)
8 The process in which plants make their own food.
11 The study of changes in weather.
14 The _____ has phases that can be seen throughout the month.
18 The body system that animals use to process food.
20 During an investigation the controlled _____ are not changed.
21 A _____ is a very large piece of land on the planet Earth.
22 The layer of air surrounding Earth.
23 A general term for when something takes place. (rhymes with sent).
25 An animal that feeds on dead things.
26 The reproductive part of a plant. (f____)
27 Tiny living things that float or drift in ocean water. (p____)
28 A mirror bounces light back so you can see your _____. (rhymes with perfection)
32 The moon casts its shadow on the Earth during this event. (2 wds.)
33 Breathing organs in mammals.
34 The hardness is a physical _____ of an object.
35 A general term for future generations of animals. (o____)

DOWN

1 The general term that describes how fast an object moves.
3 The process by which a liquid changes to a gas.
4 This liquid that can exist in 3 states and is a main part of most organisms.
5 Changes in plant species over long periods of time. (rhymes with resolution)
6 We record information in this, another term for data table.
7 Water or _____ can carry seeds away from a plant.
9 To report the results of an investigation in a concise way.
10 These are the green, food-making parts of almost every plant.
12 The only type of organisms that can create their own food energy.
13 The attraction of 2 objects to each other. (___al)
15 An organism that hunts for other animals as food.
16 A general term for something that can cover or protect a plant.
17 The type of organism that has roots that can crack and weather rocks.
19 An energy (increase decrease) changes ice into a liquid.
24 A general measurement term that measures using milliliters and liters.
29 _____ combines with hydrogen to make water.
30 The general term for push or pull.
31 These are unwanted plants in an area.

1. What force causes a volcano to erupt?

O A. Pressure from the rising temperature.

O B. Friction from the inside of the lava tube.

O C. Gravity attracting the magma to the Earth's surface.

2. What are the advantages of using a volcano model to predict volcanic eruptions?

O A. Models can show exactly how volcanoes erupt.

O B. Models can simulate how a volcano can change over time.

O C. Models allow scientists to collect data on the volcanoes changing temperatures.

3. Explain how a volcano can be both a destructive and constructive to the Earth's surface?

- Identify **one** way that a volcano can be a **destructive** to the Earth's surface.
- Identify **one** way that a volcano can be a **constructive** to the Earth's surface.

Use words, labeled pictures, and/or labeled diagrams in your answer.

Destructive force to the Earth's surface:

Constructive force to the Earth's surface:

Name: Lesson 30: Assessment 6

/ 35 = %

ACROSS

2 The ______ protects the inner organs of an animal.
4 An animal's breathing organs.
8 ______ combines with hydrogen to make water.
9 The process that changes objects from a solid to a liquid by heating.
11 Water or ______ can carry seeds away from a plant.
15 These are unwanted plants in an area.
16 There are nine ______ that orbit the sun.
18 A small, continuous flow of water.
19 An animal that feeds on dead things.
20 Changes in organisms that occur over long periods of time.
21 The luster is a physical ______ of an object.
24 Earth receives most of its _____ from the sun.
26 Producers make their own ______.
27 Another word for frozen water.
29 The term used to describe measuring the heaviness of an object.
30 To ______ is to use an object again.
31 During an investigation only one ______ is changed.
32 Producers get their energy from the ______.
33 The moon casts its shadow on the Earth during this event. (2 wds.)
34 The lungs are part of the ______ system.

DOWN

1 The liquid that carries nutrients to the different parts of the body.
3 A general term which insects, plants, and humans are example.
5 The sprouting of a plant from a seed.
6 A general measurement term that measures using milliliters and liters.
7 Ice crystals that fall in soft white flakes.
10 To report the results of an investigation in a concise way.
12 The scientific unit for measuring "a push" or "a pull."
13 We use this to represent events that cannot be easily manipulated.
14 The layer of air surrounding Earth.
17 The type of measurement used in weather and measured in degrees.
22 A general term for the steps in an investigation.
23 Seeds from plants are sometimes spread by ______.
25 An organism that hunts for other animals as food.
26 Evidence of past organism life.
28 This type of organism can give animals shelter from weather.

Parent/Guardian Review: Date:

1. What are the three states of matter?

O A. Minerals, nutrients, and vitamins

O B. Plastic, wood, and paper

O C. Solids, liquids and gases

2. How do molecules move when heated?

O A. When heated, molecules move faster.

O B. When heated, molecules move slower.

O C. When heated, molecules do not move.

3. Georgi and Elaine predicted that the chemical reaction between vinegar and baking soda would last longer if the vinegar was heated. They added 1 teaspoon of different temperatures of vinegar to ¼ teaspoon of baking soda. They timed the length of the chemical reaction and recorded these results:

Temperature of Vinegar	**Length of Reaction Trial 1**	**Length of Reaction Trial 2**	**Length of Reaction Trial 2**	**Average Length of Reaction**
40°F	93 sec.	95 sec.	92 sec.	93 sec.
70°F	59 sec.	64 sec.	62 sec.	62 sec.
100°F	30 sec.	28 sec.	31 sec.	30 sec.

Write a conclusion for Georgi and Elaine's investigation. Be sure to:

- Tell whether Georgi and Elaine's prediction was correct or not.
- Include their results of high and low data averages in your explanation.
- Compare the results of high and low data averages.

Tell whether Georgi and Elaine's prediction was correct or not:

Include their results of high and low data averages:

Compare the results of high and low data averages:

Lesson 31: "To create one's own world takes courage."
Georgia O'Keeffe, Artist

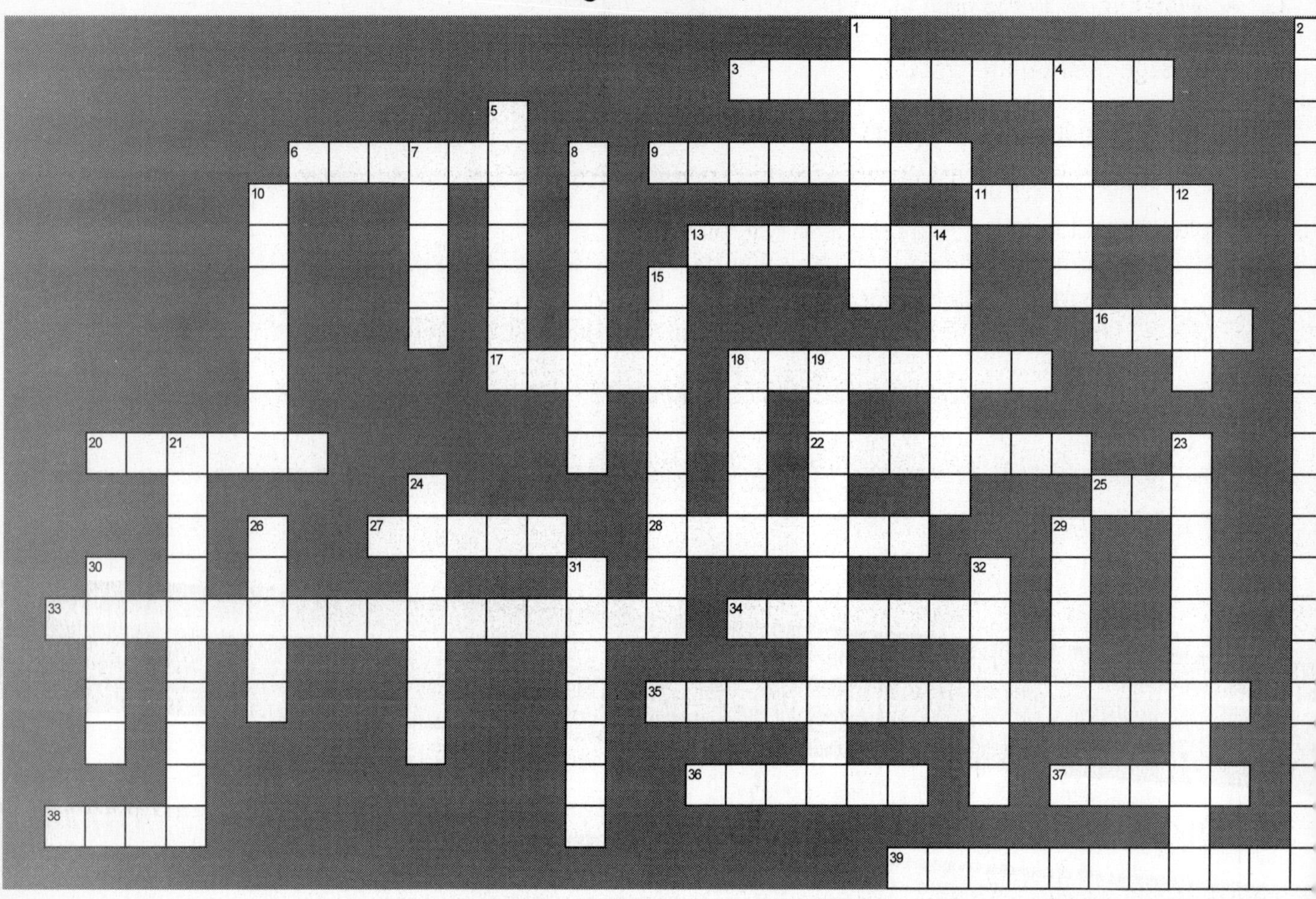

/ 40 = %

ACROSS

3 Energy from sunlight. (2 wds.)
6 An elastic tissue that move bones so humans can move. (rhymes with rustle)
9 System of bones that support the human body.
11 A general measurement term that measures using grams and kilograms. (w_____)
13 When you draw and label something, you have a _____.
16 Another term for drag or tug.
17 The earth __________on its axis.
18 Structure of bones that make up an animals shape.
20 The outcome of a change. (causes, affect)
22 Earth, Mars, Jupiter, and Venus for example.
25 The star that supplies energy to Earth.
27 All living material is made of these small units.
28 A mountain created by hardened lava.
33 Roots get ______ from the soil. (2 wds.)
34 A large body of ice that moves down a mountain.
35 This type of sound is created by slow vibrations. (3 wds.) (l_____ p_____ s_____)
36 A ________ orbits a star.
37 Structures that make up the skeletal system.
38 An animal that is hunted by another animal.
39 This occurs when pollen sticks to the end of the pistil.

DOWN

1 When objects are arrange so the next object can be predicted this is a ______. (p_____)
2 The problem or question being answered in an experiment or investigation. (2 wds.) (i_____ q_____)
4 Half the distance across a circle.
5 Another term for the outcome of an experiment. (results, energy)
7 A physical property of an object that you can see. (color, genes)
8 The general term for anything living is an ______.
10 The type of energy expressed through movement. (rhymes with athletic)
12 Thermometer, ruler and scale are all different types of ______.
14 The process of changing a solid into a liquid.
15 When you mix salt into water, the salt ______. (d_____)
18 Flowers use this property to attract pollinators.
19 The process of clearing-up a misconception. (explanation, experiments)
21 Cotyledons contain ______ for seedlings. (2 wds.) (f_____ e_____)
23 The general term for the ecosystem that surrounds an organism.
24 The general term that describes the conditions of temperature, wind speed, and precipitation.
26 This is created when an object, like a tree, blocks the sun.
29 Human breathing organs.
30 A general term for a body of water that usually starts in the mountains and flows to an ocean.
31 Aphids, crickets, and bees are examples of ______.
32 The process of changing liquid water into a solid.

1. Name one condition needed for clouds to form.

O A. Water vapor in the air.

O B. A large river on the ground.

O C. High temperatures in the atmosphere.

2. What is the source of energy needed for clouds to form?

O A. Electrical

O B. Magnetic

O C. Heat

3. Explain how clouds are formed and why it rains.

- Make a labeled diagram showing how clouds form.
- Explain why it doesn't rain every time there is a cloud in the sky.

Use words, labeled pictures, and/or labeled diagrams in your answer.

Explanation of why not every cloud causes rain:

Lesson 32: "Imagination is more important than knowledge."
Einstein, Physicist

/ 40 = %

ACROSS

2 A general term for all types of characteristic of an object.
5 Energy of motion. (2 wds.)
6 The reproductive cell of a female insect or bird.
7 A frozen, flat land with low growing plants and shrubs.
9 Write-up the final results of an investigation. (chart report)
10 To create a solution to a problem. (design reports)
11 A force is a pull or a _____.
13 _____ energy comes from the sun. (solar magnet)
16 These are small animals with 6 legs and a body divided into 3 parts.
18 The changes in the way the moon looks from Earth.
23 General weather conditions of an area over time. (c____)
26 Bike races are measured by a distance of a (foot mile yard).
27 Data or facts that proves a conclusion.
29 You put a _________ in the soil to grow a new plant.
30 The natural movement of earth materials from one place to another.
32 3 _____ = 1 yd.
33 Measuring the width of a circle is the _____.
34 A series of steps to describe how something is done.
35 The general term for leaves, straw or wood chips being placed around a plant to conserve water.
36 A type of tree that loses their leaves every year. (d____)
37 What happens to water that reaches 32° Fahrenheit?
38 What system does a human use to send messages to a muscle?

DOWN

1 The general term for the lack of water over a period of time.
3 Movement of the Earth's crust that causes the ground to shake.
4 Changing from rain to water vapor and back into rain is an example of the water _____.
8 When organisms rely on each other. (depend design)
11 The sticky part of the flower that accepts pollen.
12 The human cells that carry oxygen to all parts of the body.
14 The dusty powder produced by the stamen.
15 To arrange things in some type of order. (organize evidence)
17 The human organ used for smelling and breathing.
19 Plants can change this property as they grow.
20 A _____ is an incline plane wrapped around a center pole.
21 A small, narrow body of fresh moving water.
22 The organ that covers the body of a person.
24 A physical property that describes the surface of an object. (frozen texture)
25 The fast downhill movement of earth materials. (l____)
28 The process of adding force on an object. (pressure property)
31 A way to show data. (report, graph)
32 A push or pull is a _____.

1. What causes shadows on the earth?

O A. The sun's light is reflected by objects.

O B. The sun's light is blocked by objects.

O C. The sun's light is absorbed by objects.

2. While standing in the exact same spot on the playground, a pair of students measure and record the length of each other's shadows at 10:00 A.M., 12:00 P.M., and 2:00 P.M. What is the changed variable in this investigation?

O A. The length of their shadows.
O B. The location on the playground.
O C. The time of day

3. Janelle and Karlyn measured each other's shadows in the morning, at noon, and in the afternoon. Their results showed that their shadows were longest in the morning and afternoon, and shorter at noon.

- Explain why shadows appear longer in the morning and afternoon than they appear at noon.
- Include **a person**, the location of the **sun** in the sky, and length of the **shadows** in your drawings.
- Make sure you label your drawings as **morning**, **afternoon** and **noon**.

Use words, labeled pictures, and/or labeled diagrams in your answer.

Explain why shadows appear longer in the morning and afternoon but shorter at noon:

Draw and label a diagram in the spaces below showing why shadows are longer in the morning and afternoon and shorter at noon:		

Lesson 33: "A friend is a present you give yourself."
Robert Louis Stevenson, Author

/ 40 = %

ACROSS

4 Rocks formed from sediments. (2 wds.)
6 Another word for piece or section of an object.
8 The measurement term for the acidity level of a liquid.
9 The human cells that carry waste away from the cells.
11 To discover a new way to do something.
14 Another word for to classify or recognize an object. (identify property)
15 It frequently snows on this land formation because it is colder there.
16 A small unit of weight in the standard system.
17 What is a very large land mass?
18 Record the data into a ______. (chart lever)
20 Changes in animals species over long periods of time.
21 When liquids change to a solid state.
22 Moving ice on land that stays frozen all year.
25 To gather in crops from a garden.
26 Rocks changed by heat and pressure under the surface of the Earth.
27 A general measurement term that measures using grams and kilograms and is different than weight.
29 A community of living and nonliving things is an ______.
32 The part of the plant that holds the flowers, fruit, and leaves.
33 A large stream. (river, ocean)
35 Anything in the environment that can harm the natural resources.
36 Plants get this from the sun.
37 This is when you watch something carefully and record the data. (predictions observation)

DOWN

1 Pieces of information collected in an investigation. (part data)
2 Future generations of insects are called ______.
3 A gas exhaled by animals needed for plant survival. (2 wds.)
5 The place where the magnetic force is the strongest. (2 wds.)
7 Seeds are found inside this part of the plant.
8 A weather person can ______ tomorrow's high temperature.
10 When the Earth makes a shadow on the moon.
12 The wind moves a plant; what system moves a human?
13 Half the distance through a sphere.
19 The south poles of two magnets ______ each other.
23 An animal that is hunted for food.
24 Evidence of past plant life.
25 The human organ that pumps blood.
26 A general term for a tool that does work. (machine erosion)
28 It is the part of the flower where pollen is made.
30 ______ is a property of an object that can be sensed by animals. (rhymes with tell)
31 Hammer claw and scissors are examples of a ______. (screw lever)
34 12 inches makes one of these.

1. What is the function of the digestive system?

O A. To chew and swallow food.

O B. To provide nutrients for the cells.

O C. To remove harmful substances from the body.

2. In what organ do the digestive system and the circulatory system interconnect?

O A. Lung

O B. Brain

O C. Intestine

3. Label the parts of the digestive system and what is the function of each part.

Use words, labeled pictures, and/or labeled diagrams in your answer.

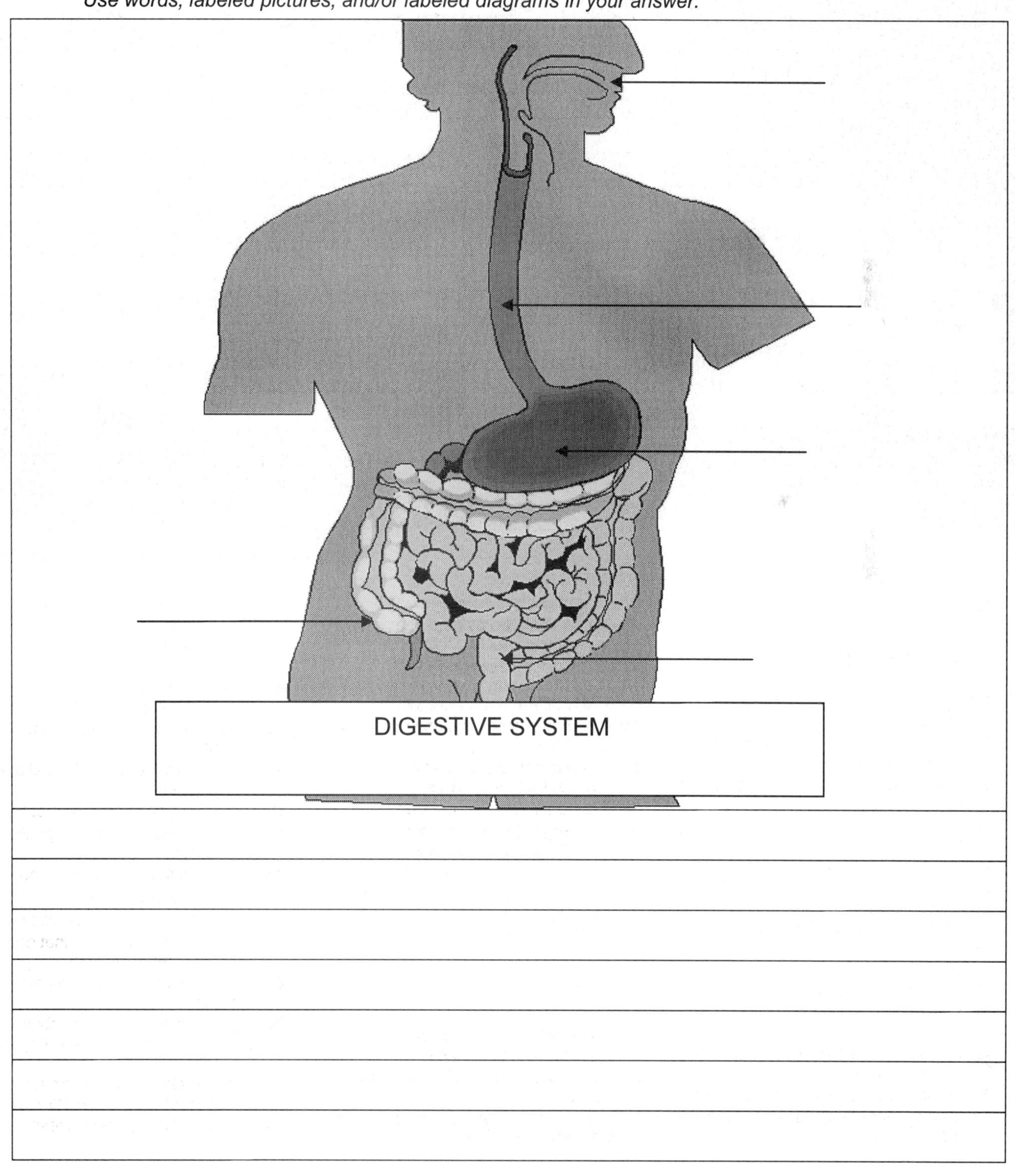

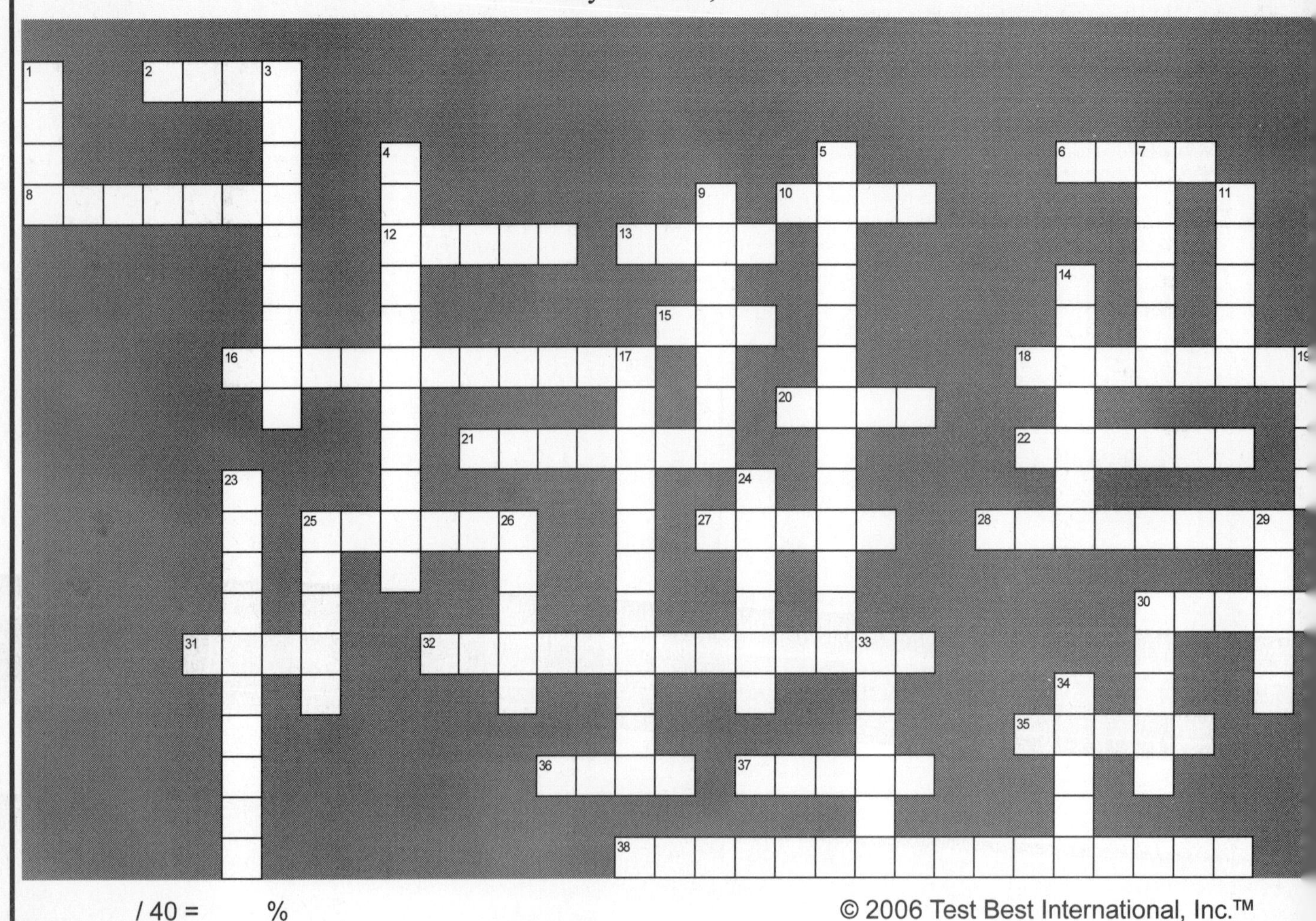

/ 40 = %

ACROSS

2 Measurement term measuring seconds, minutes and hours.
6 _____ gives all animals energy.
8 The general term for warm-blooded animals.
10 An organism that is eaten for food.
12 Systems have many _____ that work together. (rhymes with charts)
13 Drops of water falling from clouds in the form of a liquid are _____.
15 The star that supplies energy to Earth.
16 To show something. (demonstrate investigate)
18 Tiny living creatures that are food for many sea animals.
20 _____ is a physical property that can be measured by an object's volume or height.
21 The general term for a bears, insects, or humans. (habitat animals)
22 The sticky part of the flower that accepts pollen.
25 Precipitation can come from these objects.
27 These hold plants in place.
28 Any form of life is called an _____.
30 This part of the digestive system is used to take in energy.
31 This is the smallest unit of living things.
32 The force that attracts of 2 objects. (gravitational electromagnet)
35 Both a plant system and a human system need _____ to survive.
36 Seeds can be carried by water or _____ away from a plant.
37 The up and down movement of water.
38 This sound is created by fast vibrations. (3 wds.)

DOWN

1 This holds up the leaves and flowers of a plant.
3 The interaction of the organisms and the environment in a given area.
4 To clear-up a misconception. (demonstrate explanation)
5 Condensation, evaporation, and _____ are part of the water cycle.
7 Plants produce this gas.
9 Unlike solids, _____ take on the shape of their container.
11 The general term that describes adding energy into a system. (light input)
14 Another term for dragging or tugging. (p____)
17 The process by which a liquid changes to a gas.
19 The human organ used for smelling and breathing.
23 The type of energy that changes liquid water into a gas. (2 wds.)
24 Evidence of past animal life.
25 Changing from rain to water vapor and back into rain is an example of the water _____.
26 Things having to do with the Sun. (orbit, solar)
29 Metric measurement of distance similar to a yard.
30 We use this to represent systems that cannot be explored in a classroom.
33 The outcome of a change. (affect predict)
34 A very slow collection of liquid water. (lakes creek brook)

1. What is weathering?

O A. Weathering is the movement of rocks and soil.

O B. Weathering is the break down of rocks and soil.

O C. Weathering is the build up of rocks and soil.

2. What are two natural causes of weathering?

O A. Plants and ice

O B. Plants and wind

O C. Plants and fire

3. Describe two ways weathering occurs.

Use words, labeled pictures, and/or labeled diagrams in your answer.

One way weathering occurs:

Another way weathering occurs:

Name: Lesson 35: Assessment 7

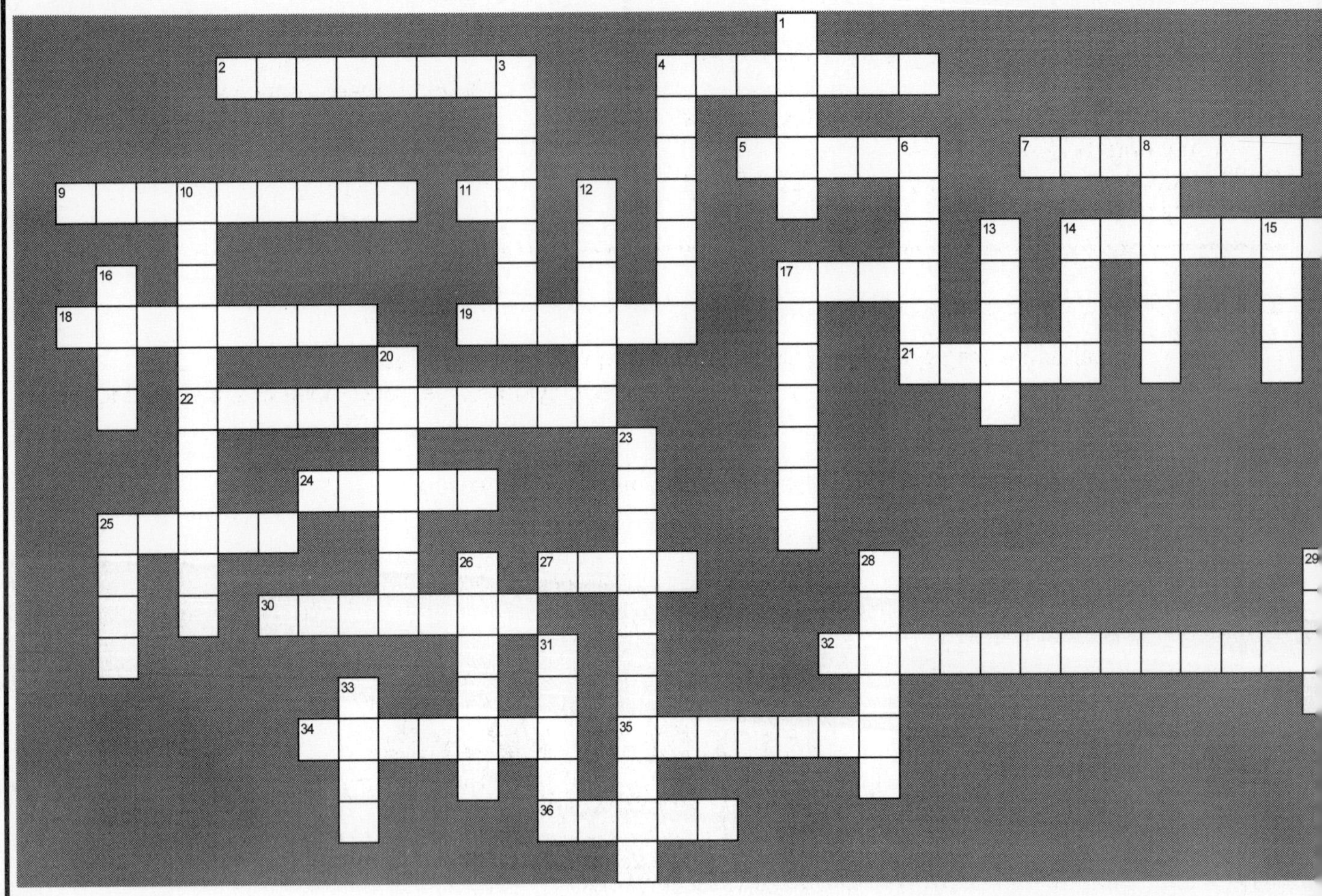

/ 40 = %

ACROSS

2 The general term for anything living is an ______.

4 Another term for dragging or tugging.

5 Structures that make up a skeleton.

7 A mountain created by hardened lava.

9 Future generations of insects are called ______.

11 The measurement term for the acidity level of a liquid.

14 A series of steps to describe how something is done.

17 Bike races are measured by a distance of a (foot mile yard).

18 To arrange things in some type of order. (organize, evidence).

19 A general measurement term that measures using grams and kilograms.

21 This part of the digestive system is used to take in energy.

22 The process by which a liquid changes to a gas.

24 ______ is a property of an object that can be sensed by animals. (rhymes with tell)

25 A push or pull is a ______.

27 This is the smallest unit of living things.

30 The natural movement of earth materials.

32 A gas produced by animals needed for plant survival. (2 wds.)

34 To gather in crops from a garden.

35 A physical property that describes the surface of an object. (frozen, texture)

36 These hold plants in place.

DOWN

1 The human cells that carry oxygen to all parts of the body.

3 A general term for a tool that does work. (machine, erosion)

4 A weather person can ______ tomorrow's high temperature.

6 A small, narrow body of fresh moving water.

8 Precipitation comes from these objects.

10 Energy from sunlight. (2 wds.)

12 Plants produce this gas.

13 The general term that describes adding energy into a system. (light, input)

14 A force is a pull or a ______.

15 The part of the plant that holds the flowers, fruit, and leaves.

16 An animal that is hunted for food.

17 The process of changing a solid into a liquid.

20 What happens to water that reaches 32° Fahrenheit?

23 The process of clearing-up a misconception. (explanation, experiments)

25 ______ gives all animals energy.

26 The dusty powder produced by the stamen.

28 A ________ orbits a star.

29 You put a __________ in the soil to grow a new plant.

31 Both a plant system and a human system need ______ to survive.

33 Pieces of information collected in an investigation. (part, data)

1. Which investigative question could be answered using only your heart rate, a timer, and increasing your activity level?

O A. How does increasing my activity level affect my pulse rate?
O B. How does increasing my activity level affect my energy?
O C. How does increasing my activity level affect my muscles?

2. Why does your heart beat faster when you are exercising?

O A. My body needs more food.

O B. My body needs more oxygen.

O C. My body needs more water.

3. Describe how blood circulates in your body.

- Label the diagram of the circulatory system to include arteries, veins, heart.
- Label the arrows to identify those that are supplying oxygen and those that are getting rid of carbon dioxide.

Use words, labeled pictures, and/or labeled diagrams in your answer.

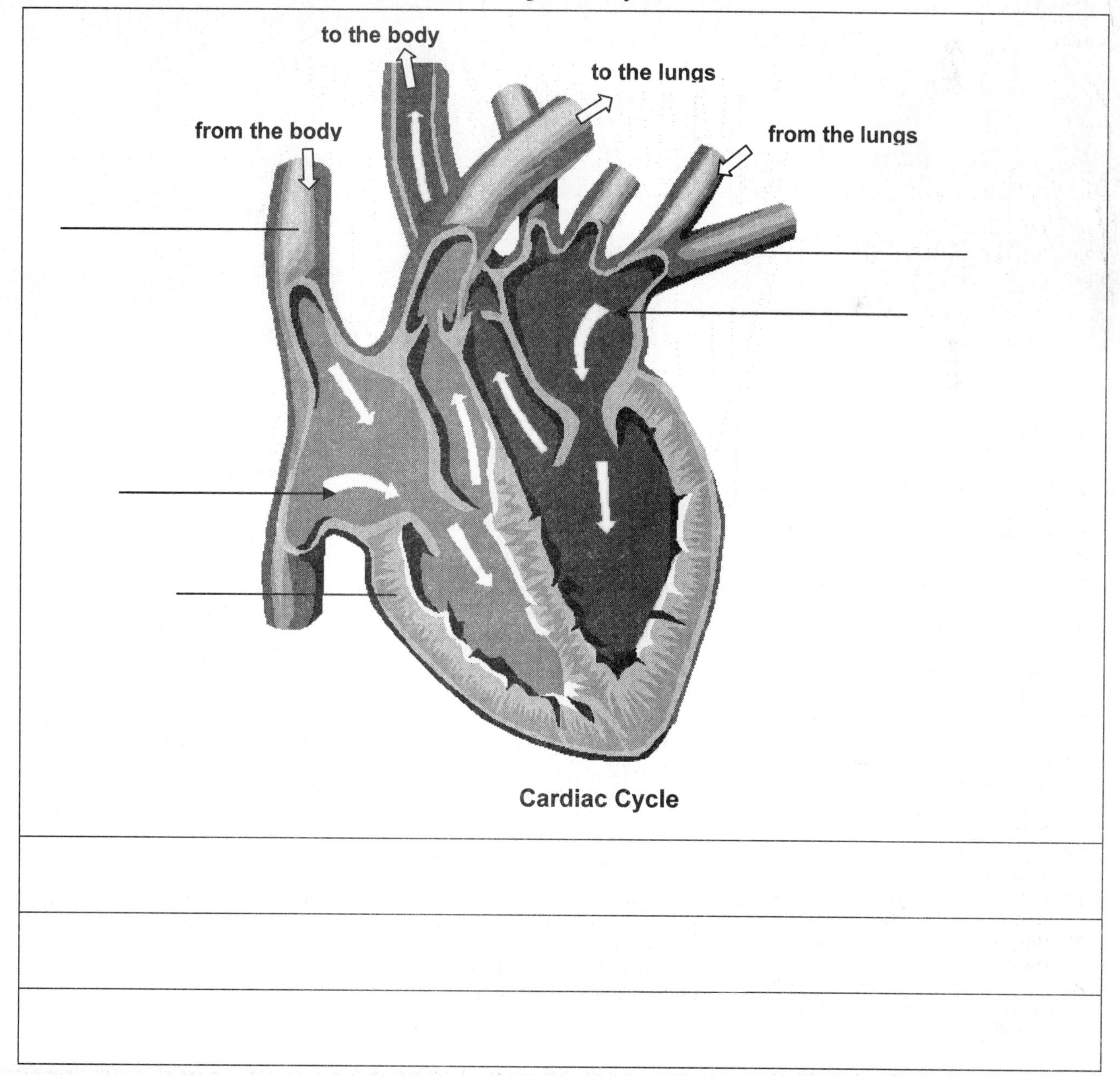

Cardiac Cycle

Lesson 36: "In order to achieve anything, you must be brave enough to fail."
Kirk Douglas, Actor

/ 40 = %

ACROSS

1 A force is a pull or a ______.
2 Ice crystals that fall in soft white flakes.
5 The liquid that carries nutrients to the different parts of the body.
6 A labeled picture showing how a system works.
7 What is a producer?
10 There are nine ______ that orbit the sun.
12 The smallest unit of living creatures. (rhymes with bell)
16 The source of energy for the water cycle.
19 Salt water that covers much of the Earth.
21 A squeezing force.
23 A mountain created by hardened lava.
24 Having to be logical and systematic. (rhymes with terrific)
29 A moving body of fresh water.
31 Seeds from plants are sometimes spread by ______.
34 The gases that surround the Earth.
35 Seedlings usually grow away from the plant to receive more light, water and ______. (2 wds.)
36 Paperclips, nails and iron filings are attracted to a ______.
37 When something repeats. (c____)
38 Earth receives most of this from the sun.
39 A measurement of length. (feet yard)

DOWN

1 Something that needs to be solved.
3 ______ combines with hydrogen to make water.
4 The cover and protection for a organism.
8 Vibrations in air or water. (rhymes with pound)
9 When you mix sugar into water and the sugar it looks like it disappears. (d____)
11 Wind and flowing water often cause this. (rhymes with explosion)
13 Melted rock that comes out of a fault in the earth.
14 Frozen water.
15 Describing how high or how low a sound is.
17 Any living thing.
18 The only organisms on Earth that can create their own food energy. (p____)
20 We use this to represent systems that are too small to be seen. (m____)
22 A lens changes the direction of light (bends light) which is called ______. (r____)
25 Humans are a ______. (producer consumer).
26 This is a unit of weight in the metric system. (rhymes with scram)
27 The act of having life.
28 When a plant species no longer exists.
30 Sound can only travel through ______ such as air, water, and solid objects.
32 The respiratory ______ is where air is taken into the body.
33 To change water from a liquid to a solid when cooled.

1. Trees are to eagles as ...?

O A. ... houses are to people.

O B. ...fish are to worms.

O C. ...dogs are to cats.

2. What is a function of a tree's roots?

O A. Absorbing energy from the ground

O B. Anchoring the tree to the soil

O C. Making food for the plant

3. How can planting trees benefit the environment?

- Identify **two** possible benefits of planting trees.
- Explain the benefit for each on the environment.

Use words, labeled pictures, and/or labeled diagrams in your answer.

One benefit of planting trees on the environment:

Another benefit of planting trees on the environment:

Lesson 37: "Never mistake motion for action."
Ernest Hemingway, Author

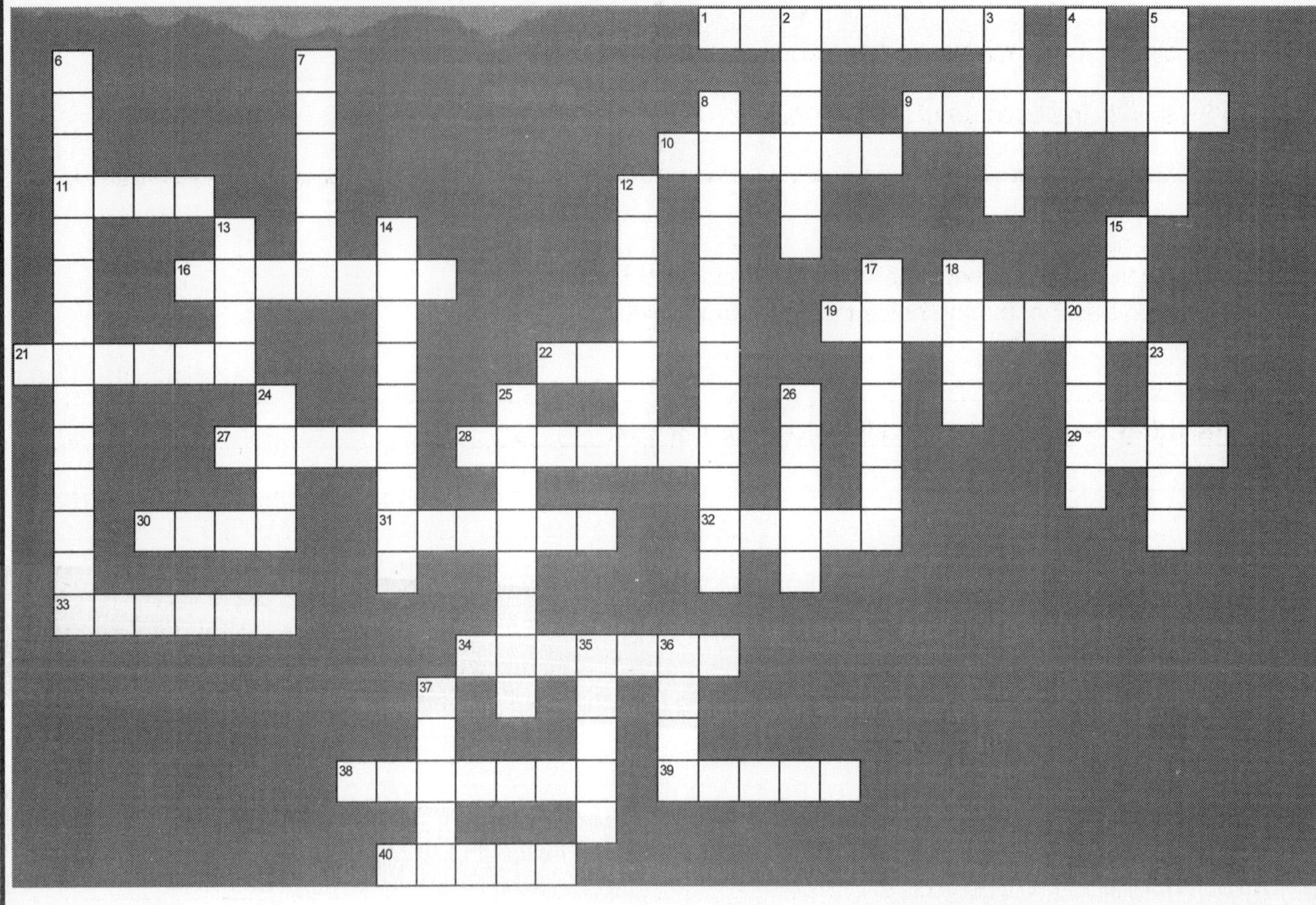

/ 40 = %

ACROSS

1 A ______ can make its own food from sunlight.
9 The luster of an object. (p_____)
10 Measured in grams and kilograms. (rhymes with late)
11 Measured in seconds, days, and centuries.
16 The movement of earth materials. (rhymes with explosion)
19 Anything that can change in an experiment.
21 Plants get this from the sun.
22 Producers get their energy from the ______.
27 This is the first part of the digestive system.
28 This is where a plant survives best.
29 We feel the Sun's light energy in the form of ______. (h_____)
30 The organ that covers the body of a person.
31 To discover a new way to do something. (rhymes with prevent)
32 A large collection of salt water.
33 Anything with parts, inputs, outputs, and a function.
34 Vertebrate animals that have hair and care for their young.
38 The lever arm rests on the ______. (f_____)
39 When a normally dry environment is under water.
40 A way to show data. (g_____)

DOWN

2 A gas found in our air.
3 The part of the plant that absorbs water and mineral nutrients from the soil.
4 Water in its solid state.
5 Over 70% of earth is covered by this.
6 This is how plants make their own food. (5 syllables)
7 ______ supply animals with oxygen.
8 The process of creating more organisms of the same kind.
12 When an animal species no longer exists.
13 An animal that is hunted and eaten by another animal.
14 An insect eats leaves, a bird eats the insect, then a hawk eats the bird is an example of a ______. (2 wds.)
15 Solid form of water.
17 We look at a ______ to make a prediction about the future. (rhymes with Saturn)
18 Some plants such as trees and grasses need ______ for pollination.
20 A form of energy produced by the sun.
23 ______ is the property that describes the form of an object. (rhymes with cape)
24 Earth's natural satellite.
25 The items used in an science demonstration. (rhymes with cereal)
26 A unit of length in the metric system.
35 Melted rock inside of a volcano.
36 The green food making part of almost every plant.
37 Energy from the sun. (s_____)

1. What is erosion?

O A. Erosion is the movement of soil and rocks.

O B. Erosion is the break down of soil and rocks.

O C. Erosion is the build up of soil and rocks.

2. What are two natural causes of erosion?

O A. Water and land

O B. Land and wind

O C. Wind and water

3. Erosion is part of an earth system.
   - Name **two** things people can do to slow down erosion.
   - Explain how each thing helps to slow down erosion.

One thing people can do to slow down erosion is:

Explain how this will slow down erosion:

Another thing people can do to slow down erosion is:

Explain how this will slow down erosion:

Lesson 38: "Whenever I don't have the answer to something, I find someone who does."
Walt Disney, Cartoonist

/ 40 = %

ACROSS

1 Organisms that eat dead organisms and waste are called ______.

9 Data that proves a conclusion. (e_____)

11 Plants take in this gas. (2 wds.)

13 A sprout is like this stage in a human's life. (rhymes with maybe)

18 Used to measure length. (inch gram)

21 The sun provides this for plants.

23 These use carbon dioxide from the air and give off oxygen. (rhymes with ants)

25 A landform surrounded by water on all sides.

28 Materials that settle out of liquids over time.

29 A body of water that usually starts in the mountains and flows to an ocean.

31 When you draw and label something, you have a ______. (d_____)

32 A large object that orbits a star.

33 Lions, sharks, and eagles are examples of ______.

34 A form of energy from the sun.

35 Uses the process of photosynthesis to make its own food.

37 When liquids like water change to a solid state.

38 Crescent, quarter, full, and new. (2 wds.)

39 The human body system that changes food into chemical energy.

DOWN

2 Something that is repeating.

3 One of the four simple properties of an object.

4 A mirror bounces light back so you can see your ______. (r_____)

5 The gasses we breathe.

6 The human system can receive energy from ______.

7 The human organ that holds and partially digests food.

8 Melted rock below the Earth's surface.

9 A community of living and nonliving things is an ______.

10 What system does a human use to send messages to a muscle?

12 When offspring receive traits from their parents. (rhymes with parrot)

14 A ______ is the opposite of a push.

15 A part of a living system inside an organism that has a specific function. (o_____)

16 Watching weather and recording is to make an ______. (rhymes with graduation)

17 The organ that allows humans to process information.

19 A ______ is a very large piece of land.

20 The study of the Earth's climate. (m_____)

22 When seeds sprout we call this ______.

24 To classify or organize. (s_____)

26 Flowers use this property to attract pollinators. (rhymes with fell)

27 A repeating series of events or actions. (rhymes with Micheal)

30 Oxygen or fruit that are produced by a plant.

36 ______ absorb water and mineral nutrients from the soil.

1. What causes a 5 pound rock to sink and a 5 pound block of wood to float?

O A. Weight
O B. Density
O C. Hardness

2. Why does a crumpled up piece of paper drop to the ground faster than a flat sheet of paper?

O A. The crumpled paper is heavier.
O B. The flat paper has more surface area.
O C. Gravity pulls more on crumpled paper.

3. When Isabel poured three liquids into a test tube she found that the liquids settled into separate layers. She tried pouring the liquids into the test tube in a different order, but again the layers ended up in the same order inside the test tube. (See diagram below.)

test tube

vegetable oil

glycerin

salt water

- Identify what is causing the liquids to form layers.
- Explain what you know about the relationship between these three liquids.

Use words, labeled pictures, and/or labeled diagrams in your answer.

Causes for the different layers:

Relationship between the different liquids:

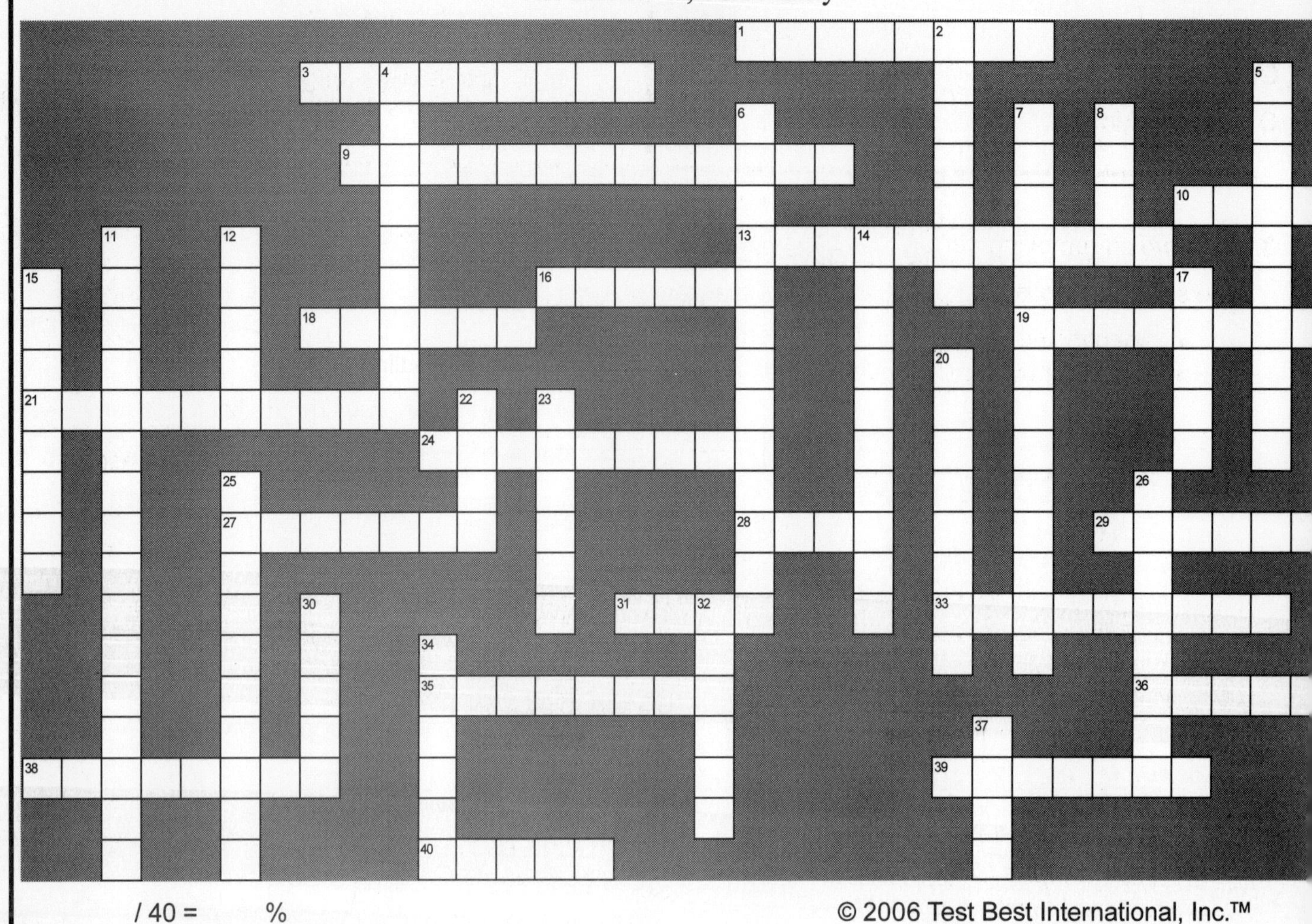

/ 40 = %

ACROSS

1 This organism eats plants or animals.
3 Steps to an investigation.
9 The scientific process of answering a question. (rhymes with communication)
10 This is what people eat.
13 To clear-up a misconception or describe an outcome.
16 Outcome of an experiment. (r_____)
18 These carry messages between animal brains and muscles.
19 System of bones that support the human body.
21 Going one way, then going another way. You are changing ______.
24 ______ makes conditions around us dirty, noisy, or unhealthy. (rhymes with solution)
27 Movement of weathered earth materials.
28 ______ forms as a result of condensation in the clouds. (rhymes with pain)
29 Something that takes up space and has mass. (rhymes with batter)
31 An animal that is hunted for food.
33 A person who studies systems.
35 This is when you arrange things in some type of order.
36 This property attracts butterflies to some flowering plants.
38 To save energy. (rhymes with preserve)
39 A large body of ice that moves down a mountain.
40 A simple machine that has a fulcrum.

DOWN

2 Tool that does work and transfers energy.
4 Bears, humans, crows and cockroaches are ______.
5 An organism that feeds on dead things.
6 Energy of motion. (2 wds.)
7 This process occurs in the leaf of a plant to supply energy.
8 Sound bouncing back to the source.
11 You find this in soil. (2 wds.)
12 We record data in this.
14 Sounds created by high energy vibrations (examples: banging objects, yelling). (2 wds.) (l_____ s_____)
15 This part of the plant contains the seeds. (2 wds.)
17 The south poles of two magnets ______ each other.
20 If you ______ the mass, then you will have more mass. (increase decrease).
22 The ______ has phases that can be seen throughout the month.
23 Where do seeds develop in a plant?
25 The type of energy that is measured with thermometer. (2 wds.)
26 A tiny bit of a substance or a molecule.
30 ______ and effect.
32 The food chain shows this moving through the ecosystem. (e_____)
34 This shows us evidence of life long ago.
37 What is an organized way to do something?

1. What is the source of energy in the baggy garden system shown below?

O A. The roots

O B. The water

O C. The sunlight

2. Which question could be investigated using **only** the materials shown in the baggy below?

O A. How does the amount of light affect the length of the roots?

O B. How does the number of seeds affect the time it takes for germination?

O C. How does the type of liquid affect the growth of the plant?

3. How would removing parts of the baggy garden system below affect the ability of the seedling to grow and be healthy?

- Identify the two **parts** of the system to remove.
- Explain how removing each of these parts would affect the seedling.

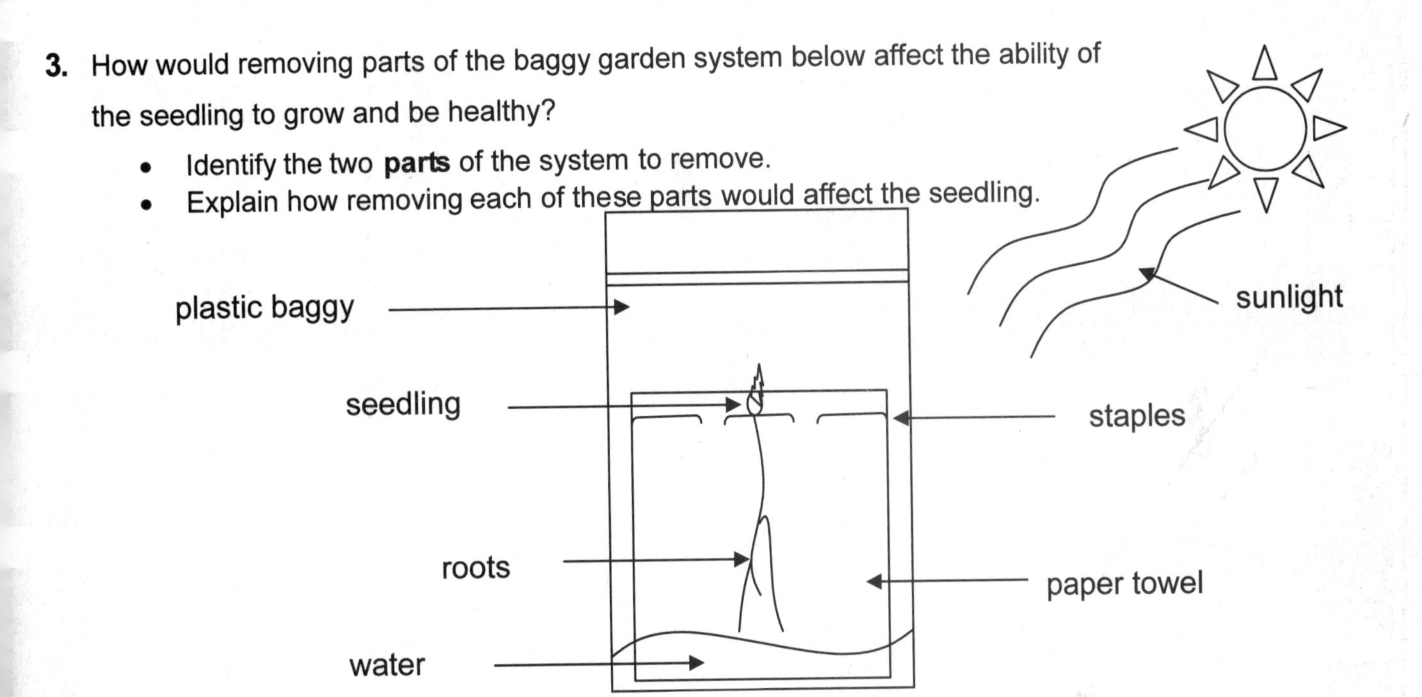

What could be removed:
How would this affect the ability of the seedling to grow and be healthy?
What could be removed:
How would this affect the ability of the seedling to grow and be healthy?

Name: Lesson 40: Assessment 8

/ 40 = %

ACROSS

2 A community of living and nonliving things is an ______.
4 The green food making part of almost every plant.
6 The items used in an science demonstration.
9 Ice crystals that fall in soft white flakes.
10 ______ combines with hydrogen to make water.
11 Lions, sharks, and eagles are examples of ______.
13 This is where a plant survives best.
14 A mirror bounces light back so you can see your ______.
16 The respiratory ______ is where air is taken into the body.
19 Sound can only travel through ______ such as air, water, and solid objects.
22 Going one way, then going another way. You are changing ______.
23 A ______ can make its own food from sunlight.
24 ______ forms as a result of condensation in the clouds.
25 A measurement of length. (feet mile)
26 A form of energy produced by the sun.
27 Used to measure length. (yard inch)
28 General term of a producer?
31 Producers get their energy from the ______.
32 We look at a ______ to make a prediction about the future.
33 Plants get this from the sun.
34 Flowers use this property to attract pollinators.
35 To save energy.
36 An animal that is hunted for food.

DOWN

1 A labeled picture showing how a system works.
3 Oxygen or fruit that are produced by a plant.
5 This is what people eat.
6 You find this in soil. (2 wds.)
7 These carry messages between animal brains and muscles.
8 Something that needs to be solved.
9 To classify or organize.
10 Salt water that covers much of the Earth.
12 This process occurs in the leaf of a plant to supply energy.
15 A repeating series of events or actions.
17 When an animal species no longer exists.
18 When seeds sprout we call this ______.
20 Outcome of an experiment.
21 Energy of motion. (2 wds.)
25 The lever arm rests on the ______.
29 Seeds from plants are sometimes spread by ______.
30 A sprout is like this stage in a human's life.

Parent/Guardian Review: Date:

1. What force causes a river to flow?

O A. Gravity

O B. Earth's water cycle

O C. Hydroelectric energy

2. Why do humans usually place dams on rivers?

O A. To control boaters and fishing on the river

O B. To improve the salmon spawning runs

O C. To generate electricity

3. Explain how organisms can affect the **flow of water** in a river system.

- Identify **two** ways that organisms could affect the river.
- Explain how these changes would affect the flow of water in the river system.

Effect on the river:
How would the flow of water be changed?
Effect on the river:
How would the flow of water be changed?

Lesson 41: "The fragrance always stays in the hand that gives the rose."
Hada Bejar, Author

/ 45 = %

ACROSS

4 ______ is a visible property of an object. (rhymes with duller)
5 The process of creating new generations.
7 To discover a new tool. (i_____)
9 When you mix salt into water it ______. (rhymes with resolves)
11 A unit for measuring "a pull". (n_____)
12 People are ______.
14 This is a plant that is beginning to grow. (rhymes with terminate)
16 Mice, insects, and rabbits eat plants, snakes eat mice, hawks eat rabbits, snakes and mice is an example of a ______.
17 The material that surrounds the roots of most plants.
20 The hardness of an object. (p_____)
21 ______ is a property of an object that can be sensed by animals through the nose.
22 This equals 16 ounces.
26 The moon changes in a predictable ______. (p_____)
27 A measurement of distance. (feet yard)
28 Comparison or interaction between objects or events. (rhymes with dealership)
31 Rocks changed by heat and pressure under the surface of the Earth. (2 wds.)
33 The red liquid in human veins and arteries.
34 The opposite of dieing.
35 An organism that hunts for other animals as food.
37 Things that never grow and reproduce.
38 The smallest part of an organism.
40 A gathering of a crop.
41 Shaft that a wheel turns on. (a_____)
42 This is a tool to see small things. (2 wds.)

DOWN

1 Pieces of information collected in an investigation. (data plan)
2 The wearing away of soil and rock.
3 Heart, lungs, skin, and brain are examples of an ______.
4 An organism that receives its energy by digesting other organisms.
5 Outcome of an investigation. (r_____)
6 The process in which plants make their own food.
8 An incline plane wrapped around a center pole.
10 Many plants grow from ______.
13 A large measurement of distance.
15 When something takes place.
18 A curved piece of clear glass that changes the direction of (bends) light.
19 Snow, rain and hail are examples of ______.
23 The colorful and fragrant part of the flower that attracts an organism.
24 Making a plan to solve a problem. (d_____)
25 ______ energy comes from electrons moving.
29 The dusty powder that is transported by insects from plant to plant.
30 Evidence of past organism life.
31 You use this to represent objects that are too small to be seen. (m_____)
32 The mixture of gasses that surround the Earth.
36 The part of the flower where seeds begin to develop. (rhymes with diary)
39 A plant part that grows on a stem.

1. What causes changes in the atmosphere?

O A. The energy from the sun.

O B. The chemicals in the lithosphere.

O C. The gravitational pull of the Earth.

2. Which is part of the hydrosphere?

O A. Volcanoes on the Earth's surface.

O B. Water vapor in the air.

O C. Plants in the river.

3. The different parts of the Earth Structure System affect each other.

- Label the parts of the Earth's Structure System using the following words.
  - ✓ Hydrosphere
  - ✓ Atmosphere
  - ✓ Lithosphere
- Explain how the hydrosphere and atmosphere affect the lithosphere.

Label the Earth's Structure System on the diagram:

Effects of hydrosphere on the lithosphere:

____

____

____

____

Effects of atmosphere on the lithosphere:

____

____

____

____

Lesson 42: "A book is like a garden carried in the pocket."
Chinese Proverb

/ 45 = %

ACROSS

1 A measurement of weight. (ounce pound)
4 Leaves, straw or wood chips can be placed around a plant for protection.
5 The Sun and everything on the Earth work together to form a weather ______. (s_____)
8 During an investigation only one ______ is changed.
12 Tool that does work. (rhymes with nineteen)
13 The study of how traits are passed from parent to offspring.
14 ______ is carried by red blood cells in human bodies.
20 This system supports your body.
21 Half the distance across a circle.
22 Changes in animals species over long periods of time. (rhymes wit revolution)
25 Receiving physical characteristics from an organism's parents. (i_____)
27 Form of an object. (rhymes with tape)
31 The tool used to measure weight. (2 wds.) (s_____ s_____)
32 Describing how high or low a sound is.
34 ______ absorbs mineral nutrients from the soil.
37 The change from a solid to a liquid by heating. (m_____)
38 Smoke is one type of air ______.
39 Any type of illness that can spread between animals and cause harm.
40 A unit to measure mass.
41 The act of being alive.
42 An experiment that tests a prediction.

DOWN

2 Effect, result, or consequence. (rhymes with paws)
3 To create a solution to a problem. (d_____)
6 You use this to represent objects that are too big to explore.
7 Steps to an experiment or investigation. ((p_____)
9 Plant systems get their energy from the ______.
10 The reproductive cell of a female of an organism like birds.
11 Light that bounces off of an object. (2 wds.) (r_____ l_____)
14 This is an item that has mass and takes up space. (rhymes with subject)
15 The part of the plant that holds the flowers, fruit, and leaves.
16 Thermometer, ruler and scale are all different types of ______.
17 To change the measurement from meters to yards. (c_____)
18 A property of the surface of an object. (t_____)
19 Decaying plant materials.
23 The part of the plant that takes in carbon dioxide from the air.
24 You put a __________ in the soil to grow a new plant.
26 When a species no longer exists.
28 An animal's organ that pumps blood.
29 The conditions of the atmosphere in a local area. rhymes with leather)
30 12 inches makes one of these.
31 This is a tool to dig a hole.
32 The dusty powder produced by the stamen.
33 This carries blood to the human heart.
35 Ice and wood are examples of this state of matter.
36 A narrow body of fresh moving water.

Parent/Guardian Signature:

For puzzle time, return on:

1. Which of the following is an example of recycling?

O A. Limiting the amount of paper supplies used.

O B. Turning newspaper into greeting cards.

O C. Re-using a cloth grocery bag.

2. Walking instead of driving is an example of which helpful activity?

O A. Reducing

O B. Re-using

O C. Recycling

3. How does recycling benefit the environment?

- Identify **two** possible benefits of recycling to the environment.

Use words, labeled pictures, and/or labeled diagrams in your answer.

One benefit to the environment:

Another benefit to the environment:

Lesson 43: "We cannot become what we need to be by remaining what we are."
Max de Pree, Author

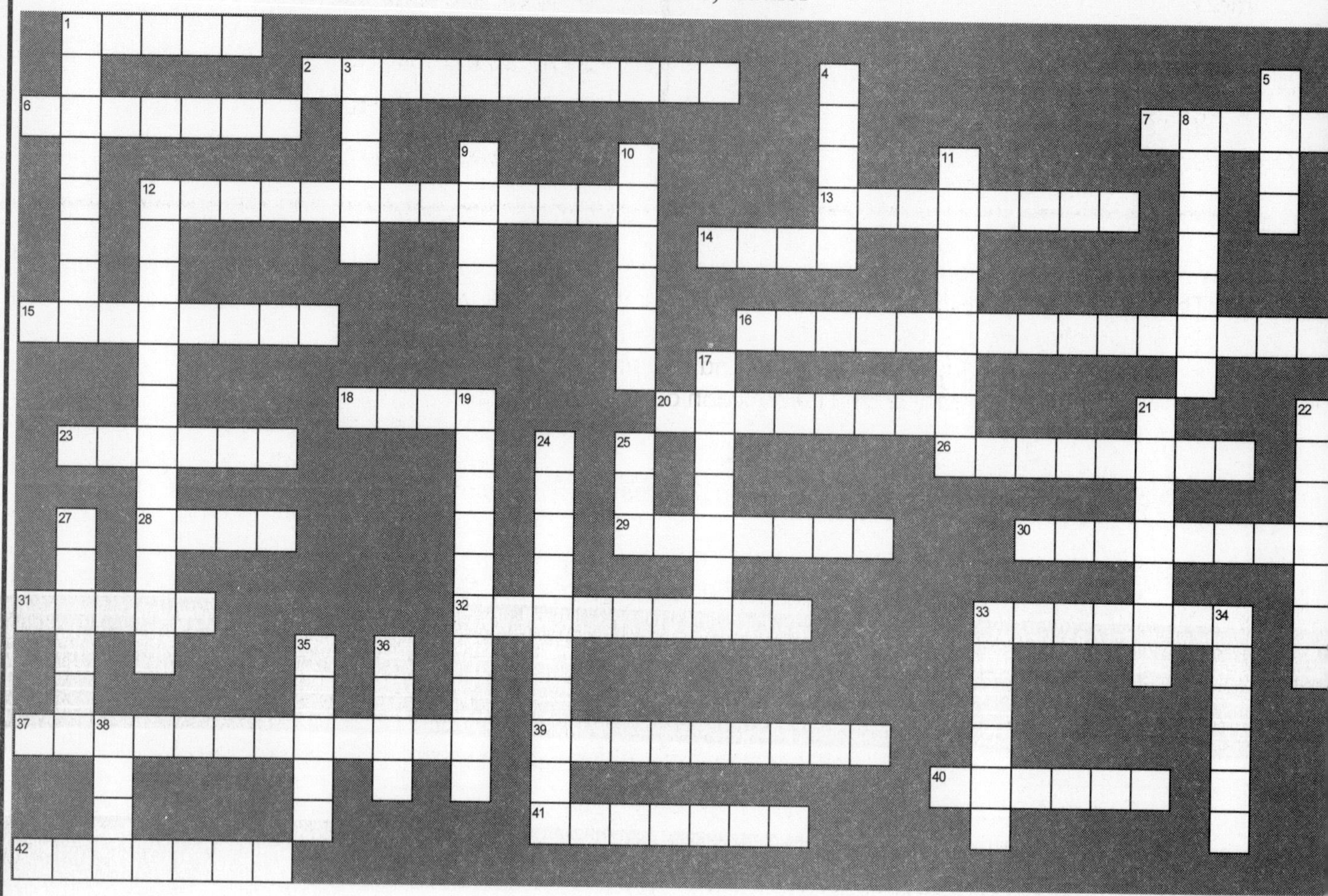

/ 45 = %

ACROSS

1 Systems have many ______ that work together. (p_____)
2 The lungs are part of the ______ system.
6 To use a substance again.
7 Plants need carbon dioxide, sunlight, and ______ for photosynthesis.
12 Plants need ______, sunlight and water for photosynthesis. (2 wds.)
13 Anything that lives is an ______.
14 Movement of air.
15 Uses the sun to do photosynthesis to create energy. (producer consumer)
16 A transparent and convex shaped lens. (2 wds.)
18 Sand and clay are examples of this. (rhymes with foil)
20 A measurement across the middle of a ball or sphere.
23 Takes in or soaks up.
26 Land that raises to great heights.
28 Measured in days, months, and years.
29 Large amount of ice moving across land.
30 Anything that can change in an investigation.
31 When the water overflows the banks of a river or stream.
32 A type of tree that has green leaves or needles all year.
33 One part of the digestive system.
37 The process by which animals create new offspring.
39 All the living things in an area.
40 A unit for measuring force or work.
41 To stake or to keep plants upright.
42 A solid natural material of only one substance.

DOWN

1 A ______ is an animal that preys on other animals.
3 When something takes place or an outcome of an investigation.
4 The human cells that fight invading bacteria.
5 3 ______ = 1 yd.
8 Organisms that move and eat other organisms.
9 The energy source for organisms.
10 Measured in ounces and pounds.
11 A natural structure on the surface of the Earth. (l_____)
12 When water vapor changes into a liquid state.
17 An illness that can spread between plants and cause harm.
19 The flash of light during a rainstorm.
21 A place where plants live.
22 These are small animals with 6 legs and a body divided into 3 parts.
24 Hyenas and vultures are ______.
25 A reproductive cell of an organism.
27 Measured in liters. (v_____)
33 A digging tool that is also a lever.
34 Measured in inches, feet, and yards. (rhymes with might)
35 A series of events that happen over and over again.
36 ______ is a liquid form of precipitation.
38 Steps to complete an investigation or experiment.

Parent/Guardian Signature:

For puzzle time, return on:

1 Mary found a rock with tiny bits of sea shells in her backyard. What is the classification of this rock?

O A. Metamorphic rock

O B. Igneous rock

O C. Sedimentary rock

2 Which variable is investigated if the same size and type of mineral is placed in water, orange juice, and soda pop?

O A. Size of mineral

O B. Type of container

O C. Type of liquid

3 There are three classifications of rocks; igneous, sedimentary, and metamorphic.

- Explain how igneous, sedimentary, and metamorphic rocks are formed.
- Give one example for each classification of rock.

Use words, labeled pictures, and/or labeled diagrams in your answer.

How are igneous rocks formed?
Example of an igneous rock:
How are sedimentary rocks formed?
Example of a sedimentary rock:
How are metamorphic rocks formed?
Example of a metamorphic rock:

Lesson 44: "What the heart knows today the head will understand tomorrow."
James Stephens, Author

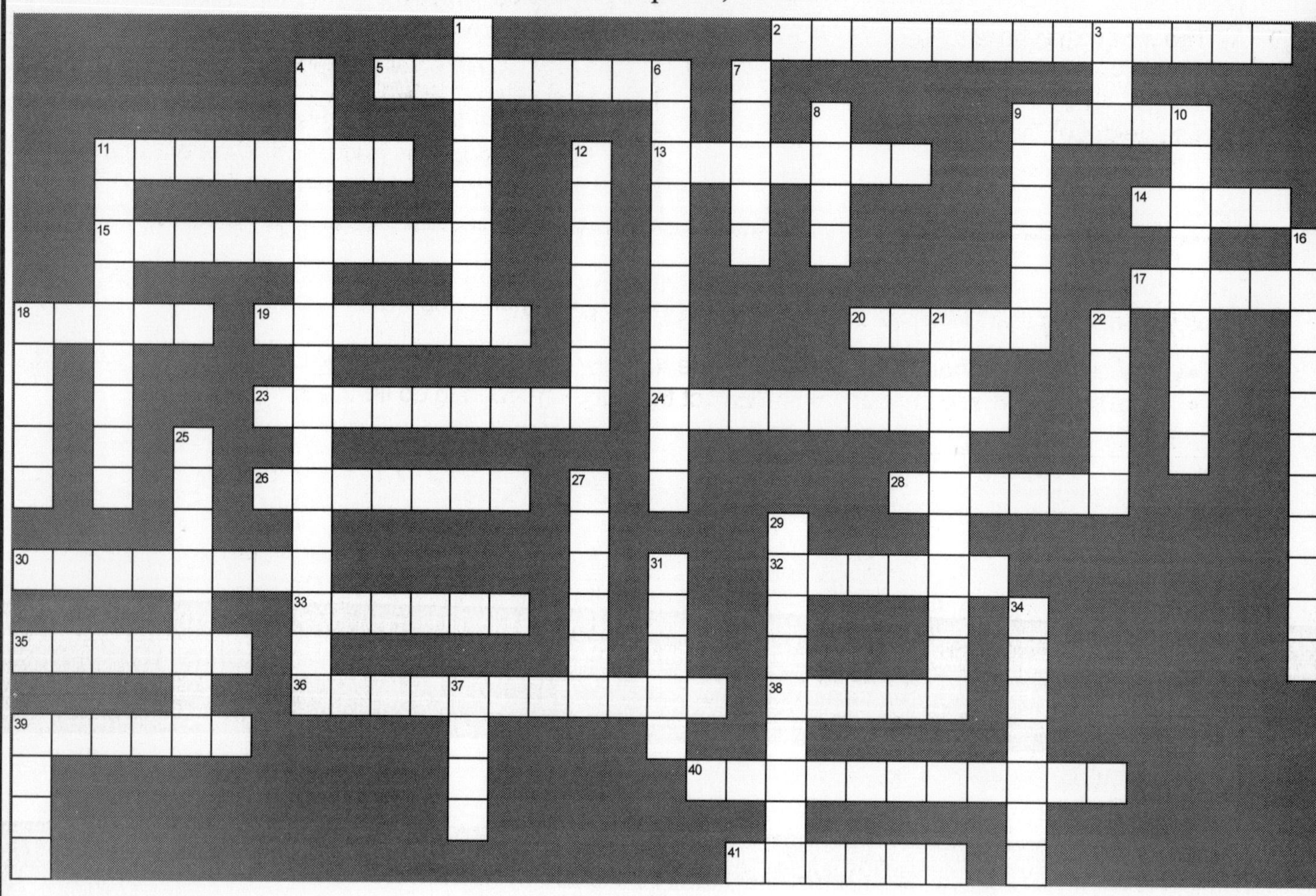

/ 45 = %

ACROSS

2 When pollen is transferred from the pistil to the stamen of the flower.
5 Uses the process of photosynthesis to create energy. (consumer producer)
9 A push or pull is a ______.
11 Measured in centimeters, meters, and kilometers.
13 Changing the amount of light, temperature and weather throughout the year. (s_____)
14 The natural satellite seen from Earth.
15 To change the measurement from liters to quarts.
17 A ______ is an example of an omnivore. (human eagle)
18 Two incline planes back to back.
19 Moving ice on land that stays frozen all year. (g_____)
20 The life ______
23 Another word for material. (s_____)
24 Future generations of animals. (o_____)
26 To make a guess about the outcome of an event.
28 A substance that can flow freely. (rhymes with fluid)
30 A ______ is anything that can change in an investigation.
32 Humans breathe this gas to function.
33 The sweet smelling part of a flower that attracts pollinators.
35 Evidence of past plant life.
36 The Sun and all the objects that orbit the sun. (2 wds.)
38 All living material is made of these.
39 The digestive ______ is a process of breaking down food for energy.
40 When liquid water changes into water vapor.
41 The food chain shows ______ moving through the ecosystem.

DOWN

1 To change water vapor into liquid water.
3 The gas surrounding Earth.
4 Roots get ______ from the soil. (2 wds.)
6 The ______ system provides the body with oxygen.
7 The roots of this organism can crack and weather rocks.
8 Plants can make their own ______.
9 The process of changing liquid water into a solid.
10 Changes in organisms that occur over long periods of time. (rhymes with noise pollution)
11 Trees that loose their leaves in the fall are called _____ trees.
12 This is a visual aide in a diagram. (p_____)
16 The ecosystem that surrounds an organism. (e_____)
18 These are unwanted plants in an area.
21 There are 7 of these large land forms.
22 Energy made when objects vibrate. (rhymes with found)
25 To arrange from largest to smallest for example.
27 Photosynthesis occurs in the plant's ______.
29 The process of getting food from one organism to another. (2 wds.)
31 Liquid that can exist in 3 states and is a main part of most organisms.
34 The force that causes rain to fall to Earth.
37 The ______ cycle is a process in which rocks are formed and changed.
39 The location of mineral nutrients for plants.

Parent/Guardian Signature:

For puzzle time, return on:

1 What is causing the lever pictured below to drop down at one end?

O A. The size of the boxes

O B. The weight of the boxes

O C. The height of the fulcrum

2 Which investigative question could **not** be answered using only the materials shown in the system below?

O A. How does moving the fulcrum affect the lever system forces?

O B. How does the height of the fulcrum affect the lever force system?

O C. How does adding more weight to the lever arm affect the lever system forces?

3 Explain two ways that the small box could be used to lift the larger box.

- Identify **two** changes that could be made to the system to allow the 10 lb box to lift the 30 lb box.
- Explain how these changes would affect the forces needed to do lift the 30 lb box.

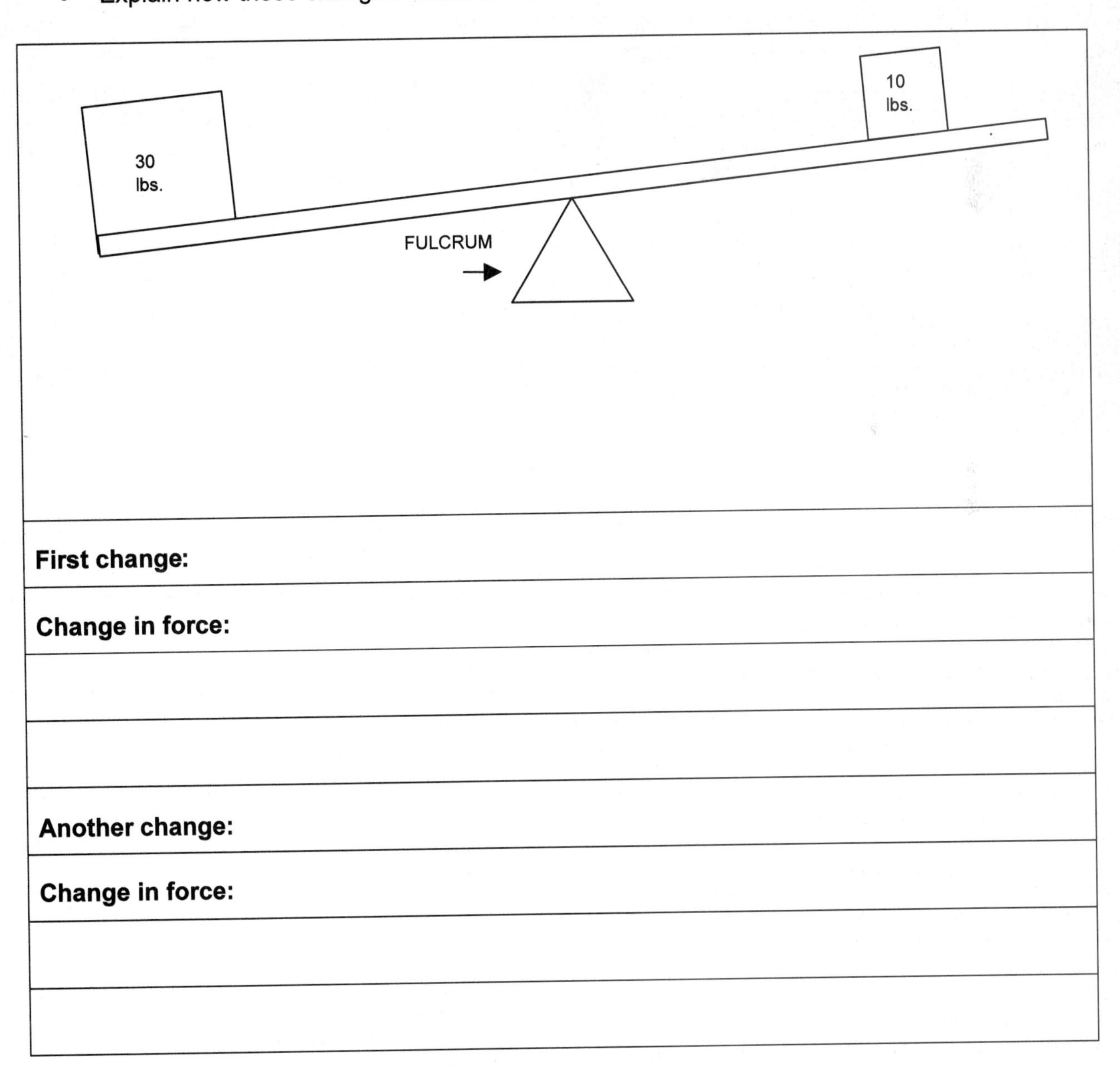

Name:

/ 45 = %

ACROSS

2 During an investigation only one ______ is changed.
3 A measurement across the middle of a ball or sphere.
7 The dusty powder produced by the stamen.
9 The conditions of the atmosphere in a local area.
11 A unit for measuring "a pull".
13 The material that surrounds the roots of most plants.
14 Organisms that move and eat other organisms.
15 One part of the digestive system.
16 The colorful and fragrant part of the flower that attracts an organism.
17 Leaves, straw or wood chips can be placed around a plant for protection.
20 These are unwanted plants in an area.
22 The hardness of an object.
23 ______ is carried by red blood cells in human bodies.
24 Plant systems get their energy from the ______.
25 This is a visual aide in a diagram.
29 The reproductive cell of a female of an organism like birds.
30 The force that causes rain to fall to Earth.
34 The food chain shows ______ moving through the ecosystem.
35 A substance that can flow freely.
36 Movement of air.
38 12 inches makes one of these.
40 When liquid water changes into water vapor.
41 To discover a new tool.
42 Land that raises to great heights.
43 The opposite of dieing.

DOWN

1 Measured in days, months, and years.
4 When something takes place or an outcome of an investigation.
5 All living material is made of these.
6 An organism that receives its energy by digesting other organisms.
8 The flash of light during a rainstorm.
10 The ______ system provides the body with oxygen.
12 Decaying plant materials.
13 Form of an object.
14 Shaft that a wheel turns on.
18 The Sun and everything on the Earth work together to form a weather ______.
19 To change the measurement from liters to quarts.
21 The Sun and all the objects that orbit the sun. (2 wds.)
26 Changes in organisms that occur over long periods of time.
27 An incline plane wrapped around a center pole.
28 Heart, lungs, skin, and brain are examples of an ______.
31 These are small animals with 6 legs and a body divided into 3 parts.
32 When the water overflows the banks of a river or stream.
33 A measurement of distance.
37 An animal's organ that pumps blood.
39 This equals 16 ounces.

Parent/Guardian Review: Date:

1. Which of the following materials should be placed in a compost bin?

O A. Plastic bags

O B. Leaves

O C. Cans

2. What is the role of worms in a compost bin system?

O A. Producers

O B. Consumers

O C. Decomposers

3. Why is composting helpful to the environment?

- Identify **two** possible benefits of composting to the environment.

Use words, labeled pictures, and/or labeled diagrams in your answer.

One benefit to the environment:
Another benefit to the environment:

Graph Your Progress

Record your personal goal for each puzzle, then shade in each line of the graph to show the number correct.

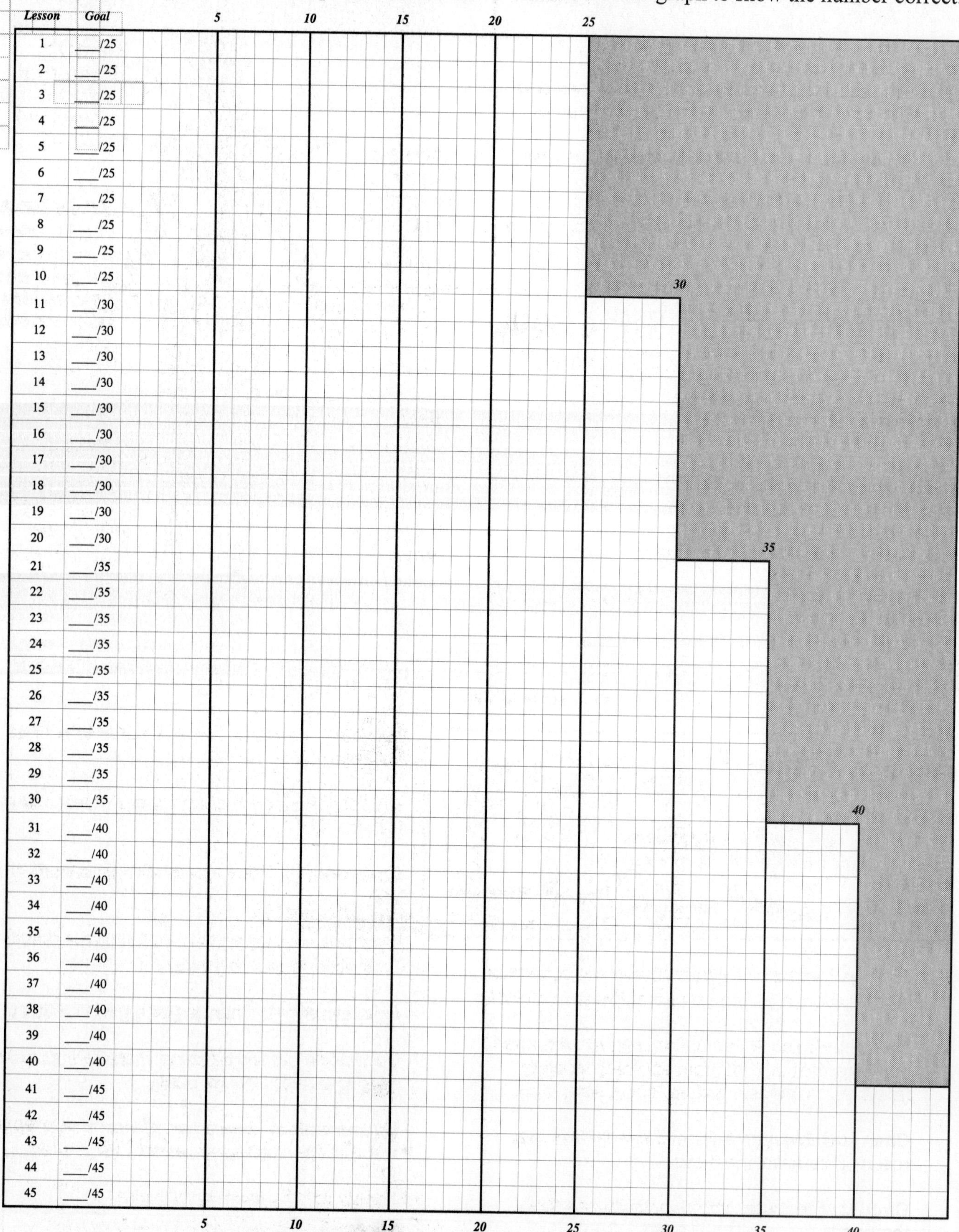

Glossary

Absorb: To take in.

Air: Clear odorless gas surrounding the Earth.

Amount: The quantity of something.

Animals: Any living organism that moves and eats other things, also known as a consumer.

Atmosphere: Layer of air that surrounds the Earth.

Axis: An imaginary line that passes through an object around which the object seems to turn.

Balance: A tool to measure mass, also known as a balance scale.

Brain: An organ in the nervous system that controls the body.

Carbon dioxide: Gas from the air that plants use to make their own food.

Cause: Results of something.

Carnivore: Animals that eat other animals, like a tiger or alligator.

Cell: Smallest unit of life that can grow.

Plant Cell & **Animal Cell**

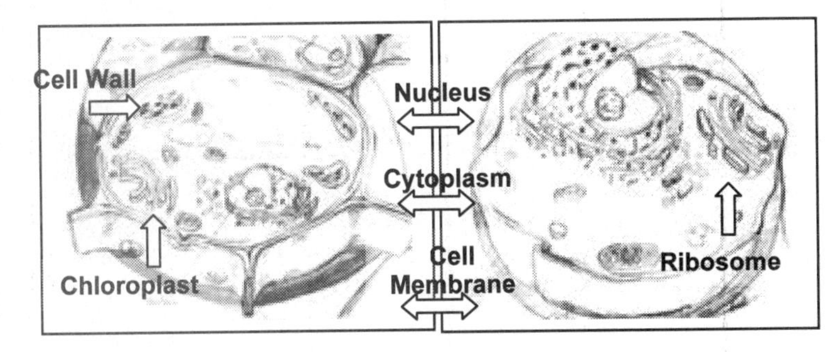
Cell Wall
Nucleus
Cytoplasm
Cell Membrane
Chloroplast
Ribosome

Changed variable: What is being changed in an investigation, also known as manipulated variable.

Characteristic: A trait or feature that sets apart one organism or thing from another; property (example: hardness, texture, color, size, etc.).

Chemical Energy: A form of energy that is a result of chemical reactions.

Circuit: The route which electricity can flow through.

Circulatory System: The system that moves blood through the body.

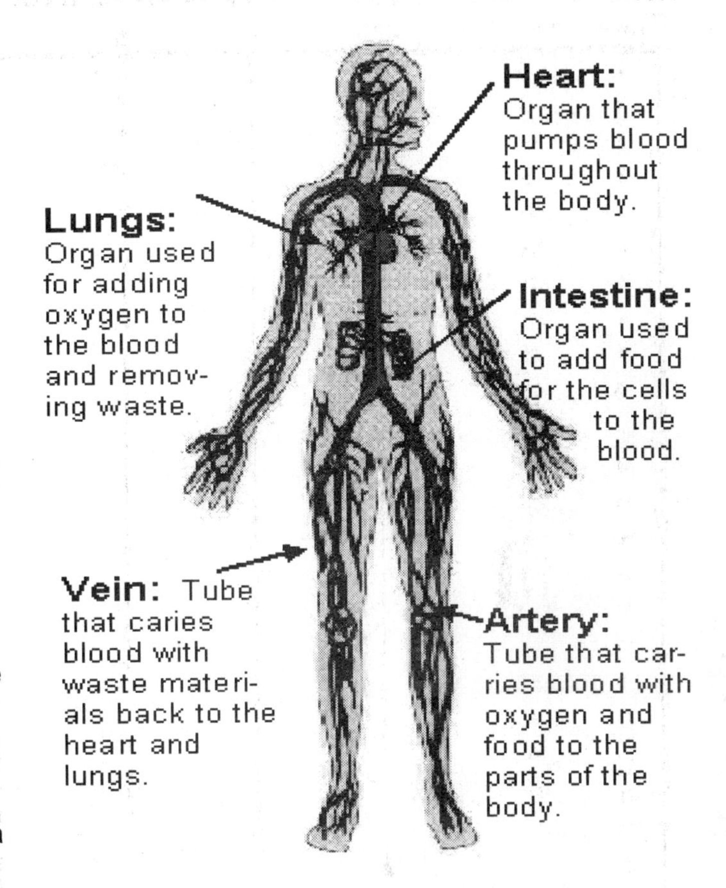
Heart: Organ that pumps blood throughout the body.
Lungs: Organ used for adding oxygen to the blood and removing waste.
Intestine: Organ used to add food for the cells to the blood.
Vein: Tube that caries blood with waste materials back to the heart and lungs.
Artery: Tube that carries blood with oxygen and food to the parts of the body.

Classify: A grouping of organisms or objects based on characteristics.

Climate: A pattern of weather over time in an area.

Cloud: A collection of condensed water vapor that is in the air.

Composting: The process of making nutrient rich soil.

Conclusion: A statement that tells the results of an investigation using data.

Condensation: Changing of a gas to a liquid.

Conductor: A substance or material that allows energy to pass through easily.

Conservation: To protect, preserve or keep from harm; things such as air, water, trees, energy.

Construct: To build up; to make.

Glossary

Consumer: An organism that eats another organism to get energy; herbivore, carnivore, or omnivore

Core: The inner most layer of the Earth, the center of the Earth.

Cotyledons: The part of the seed that contains food energy for the plant until it can make its own food; seed leaf.

Crust: The outer most layer of the Earth.

Cycle: When something happens over and over again.

Data: Facts, pieces of information, numbers or words collected during an investigation.

Deciduous: Trees that lose their leaves in the fall.

Decomposer: An organism that eats dead material and wastes to get energy.

Desert: An area with very little rainfall or water, and few plants.

Destructive: To break down, wreck or destroy.

Digestive System: The system that breaks down food for the body to use.

Earthquake: A sudden movement of the Earth's crust.

Eclipse: Blocking the sunlight from one object to another.

Ecosystem: Living and nonliving things in an area or habitat.

Echo: A sound that is repeated because the sound bounces around.

Electrical Energy: A type of energy from electrons moving.

Energy: Being able to do work.

Energy Forms: Heat, light, sound, electrical, chemical, nuclear, kinetic and potential.

Energy of motion: Kinetic energy.

Environment: Where an organism lives and reproduces.

Erosion: The process when soil or rock is moved from one place to another. This is not weathering.

Evaporation: Changing from a liquid to a gas.

Evergreen: Plant that remains green all year.

Illustration: the **Douglas Fir** tree is a conifer (has cones) and is an **evergreen** tree.

Fertilizer: Substance that is added to soil to provide mineral nutrients for plants.

Food: Nutrients produced by plants or something that is consumed by animals to supply energy.

Food Chain: Showing the food energy path from one organism to another organism.

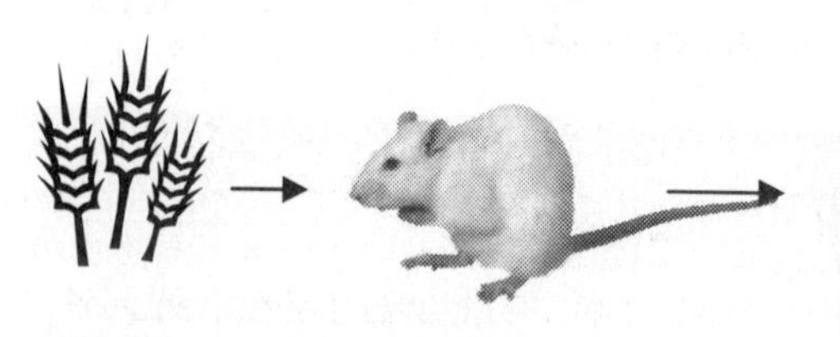

Food Web: All the different food chains in an ecosystem.

Force: To move, a push or pull.

Forest: An area with lots of trees and other plants.

Fossil: Evidence of an organism from the past.

Fossil remains: Fossils can be imprints, bones changed into minerals, footprints, etc.

Freeze: To change from a liquid to a solid due to low temperatures.

Fruit: That which grows from a flower and contains the seed(s).

Fulcrum: The pivot point of a lever system; a part of a simple machine.

Function: The purpose or reason for something.

Glossary

Gas: A substance that has no definite volume or shape, air.

Germination: Plants beginning to grow or sprout from a seed.

Glacier: A huge mass of ice slowly moving over the land.

Grassland: A prairie, meadow or other large grassy place.

Gravity: A force that pulls things together.

Habitat: An area where an organism usually lives; an environment.

Hand Lens: A tool to make things look larger; magnifying glass.

Harvest: To gather crops.

Heat Energy: Energy from moving particles; thermal energy.

Herbivore: An organism that eats plants; cow or rabbit, mouse, deer.

Humidity: The amount of water vapor in the air.

Hydrosphere: All the water on the Earth's surface and in the air.

Hypothesis: An educated guess with a reason.

Inclined plane: A flat object on a slant; a simple machine.

Igneous rock: Rocks formed when magma cools and hardens.

Inherit: To receive a characteristic or trait from the parents.

Input: An addition to a system or substance; An action or force that causes a system to work. (transfer of matter or energy)

Insects: Organisms with six legs and three body sections (Head, Thorax, & Abdomen).

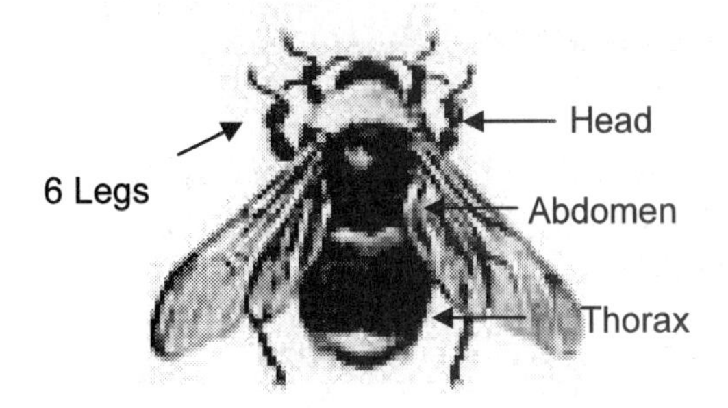

Insulator: A substance or material that does not allow energy to pass through easily; rubber, wood.

Investigation: To try and answer a question by using the scientific process.

Question: Ask a question that can be tested.

Prediction: A guess about what the answer to the question will be.

Materials: Things needed to answer the question.
Procedure: Steps to answer the question

Variable changed (manipulated): the one thing that is being tested or asked about in the question.

Measured variable (responding): How the investigation will be measured so that the question can be answered.

Variable kept the same (controlled): Everything else in the investigation that needs to stay the same so the question can be answered.

Data: Writing down what happened when doing the investigation.

Conclusion: The results of the investigation.

Lake: A large, inland body of water.

Lava: Melted rock found on the Earth's surface. (See volcano.)

Leaf: The place where a plant makes food using carbon dioxide, water, and the sun's energy. (See also plant & photosynthesis.)

Glossary

Lever: A bar on a fixed point that can be pushed or pulled on; a simple machine.

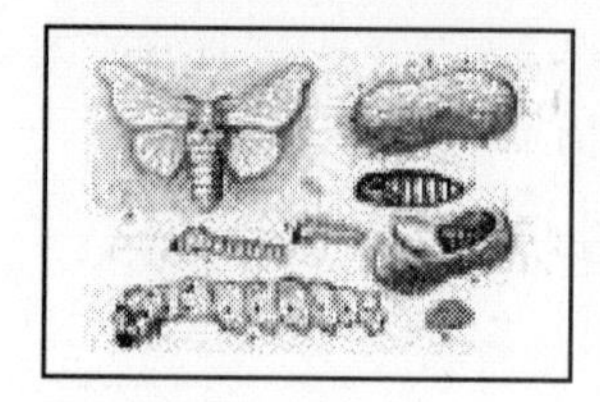

Life Cycle: The stages an organism goes through during its life. For example: seeds become plants, plants grow flowers, and flowers make seeds.

Light Energy: A form of energy that can be seen by the eye.

Lithosphere: Outer layer of the Earth including the crust and the part of the mantle.

Living: Being alive; growing, reproducing, breathing, eating.

Liquid: A substance that has a definite volume but no definite shape; water.

Machine: A tool that makes work easier.

Magnet: An object that attracts steel and iron. Magnets can attract or repel other magnets.

Magnifying Glass: A tool to make things look larger; hand lens.

Magma: Melted rocks beneath the Earth's surface. (See volcano.)

Mass: The amount of matter or stuff in an object.

Material: What something is made of.

Material List: A list of items you need to carry out an investigation.

Melt: To change from a solid to a liquid.

Metamorphic rock: Rocks that are made through heat and pressure.

Mineral nutrients: Substance that plants need from the soil to grow.

Model: An idea, picture or object that represents or shows how a system works.

Mountain: A part of land that is higher than the area around it; a very high hill.

Moon: The "natural satellite" that moves around the Earth and displays different phases throughout the month.

Mulch: Material such as leaves, straw, and bark that can hold moisture in the soil.

Muscle: The tissue that makes the parts of the body move or function.

Nectar: Liquid inside the flower that attracts pollinators.

Nerve: A system of cells that carries signals back and forth between the brain and the parts of the body.

Nervous System: The system that carries signals back and forth between the brain and the parts of the body.

Non-living: Never having been alive; examples such as rocks and air.

Nutrients: Substances that organisms need to grow and survive.

Observation: To write down what can be seen, smelled, touched or heard. **Observe:** To study carefully.

Ocean: Large body of salt water that covers most of the Earth.

Omnivore: An organism that eats plants and organisms; example: people or bears.

Orbit: The path an object travels around another object.

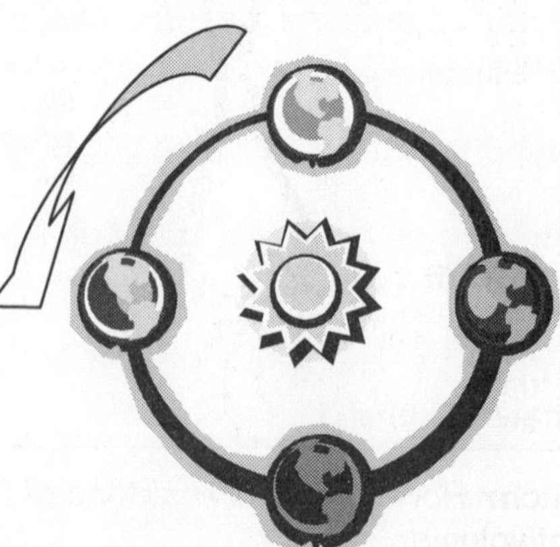

Organism: A living thing.

OR

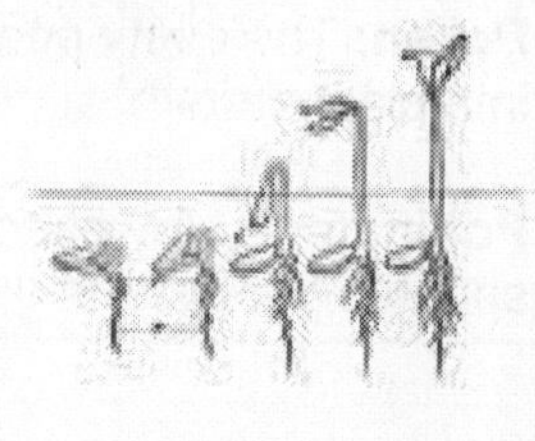

Glossary

Output: An action or force that is the result of a system at work (transfer of matter or energy).

Oxygen: Colorless, odorless gas that is found in the air that animals breathe.

Pesticide: A substance to control pests.

Photosynthesis: A plant process that occurs in the leaf using sunlight, air (carbon dioxide), and water to produce food for the plant.

Pistil: The sticky part of the flower that accepts pollen.

Planet: A large body that orbits a star like the sun.

Plant: An organism that makes food using the sun's energy; producer.

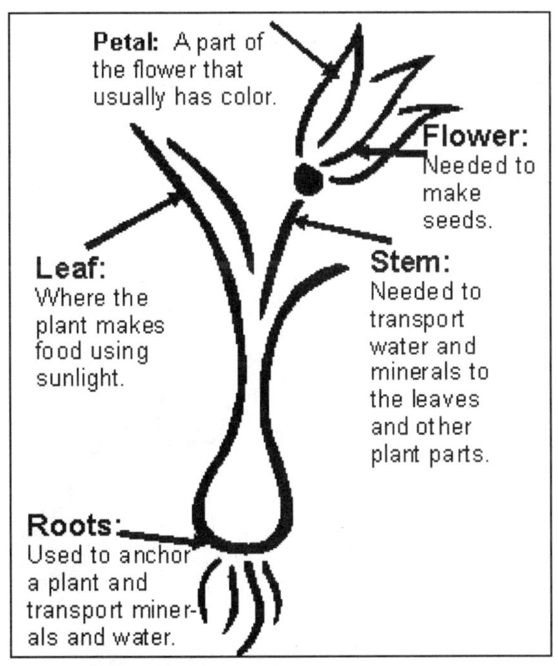
Petal: A part of the flower that usually has color.
Flower: Needed to make seeds.
Leaf: Where the plant makes food using sunlight.
Stem: Needed to transport water and minerals to the leaves and other plant parts.
Roots: Used to anchor a plant and transport miner-als and water.

Pitch: How high or low the tone of a sound is; not volume.

Pollen: The dusty powder produced by the stamen in a plant system.

Pollinators: An insect or other organism that spreads pollen from flower to flower.

Precipitation: Water that falls to the Earth; rain, snow, sleet, and hail.

Predator: An organism that hunts and eats other organisms; examples: lion or shark.

Pressure: The force that is acting over any given surface or area.

Prey: An organism that is hunted or eaten by predators; mouse or rabbit.

Producer: An organism that uses sunlight to make its own food; plant.

Properties: A trait or feature that tells apart one organism or thing from another, characteristic. (Example: hardness, texture, color, size, etc.)

Pull: To give force to an object so that the object moves toward the force.

Pulley: A wheel with a rope that can move a heavy load; a simple machine.

Push: To give force to an object so that the object moves away from the force.

Recycle: Reusing materials to make new materials.

Respiratory System: The system that inhales oxygen into the lungs and exhales carbon dioxide.

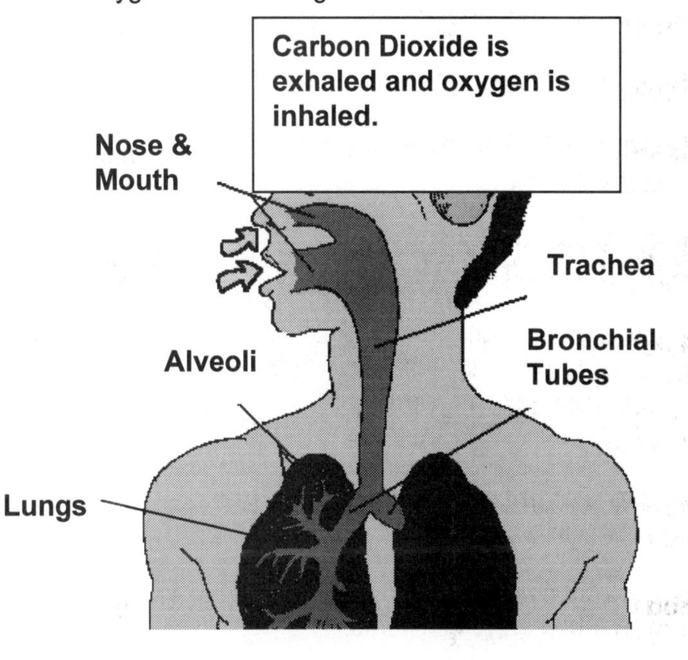
Carbon Dioxide is exhaled and oxygen is inhaled.
Nose & Mouth
Trachea
Bronchial Tubes
Alveoli
Lungs

Glossary

River: A large natural stream usually emptying into a lake, ocean, sea or another river.

Saturated: To soak or fill to capacity.

Scavenger: An organism that feeds on decaying or dead organisms; vulture.

Screw: An inclined plane that twists around a center point; a simple machine.

Sea: A salty body of water which is not as large as an ocean.

Sedimentary rock: Rock formed from layers of sand, clay and other sediments.

Seed: New plants grow from this part of the plant.

Seed pod: Part of the plant holds the seeds.

Seeds

Seed Pods

Shelter: To cover or protect something.

Simple Machines: Tools that have few or no moving parts and make work easier: inclined plane, lever, pulley, wheel & axle, screw, wedge.

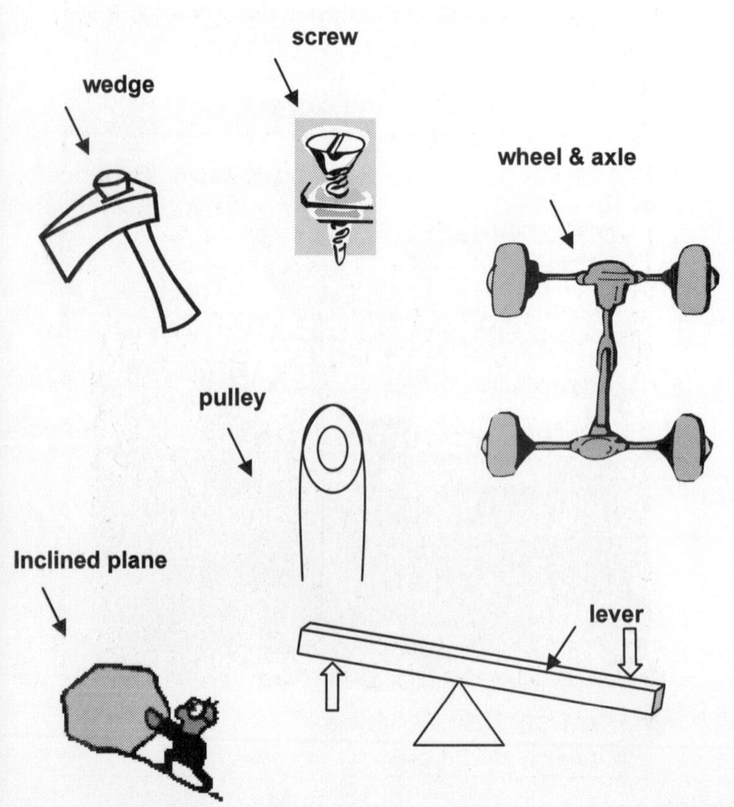

Skeletal system: The structure of bones that supports the body.

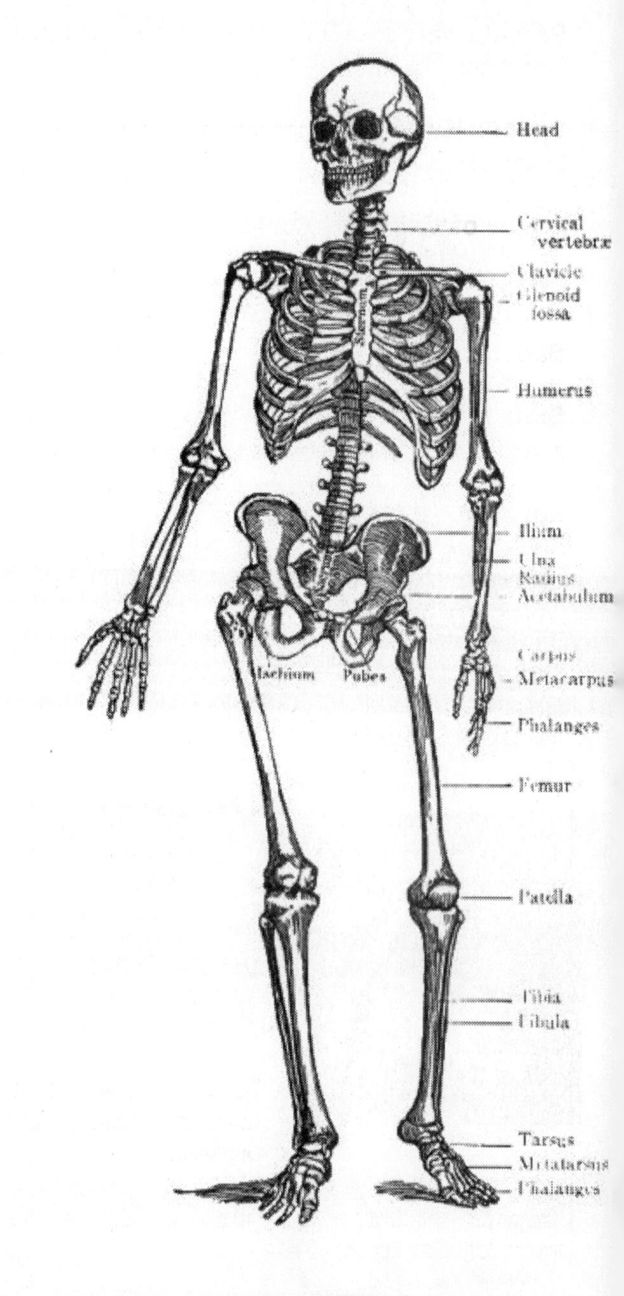

Skin: Outside layer of tissue that protects the body; largest organ of the body.

Soil: Dirt or a mixture of clay, sand, mulch, mineral nutrients, etc.

Solar system: The sun and everything that moves around the sun including planets, moons, asteroids, comets, etc.

Solid: A substance that has a definite volume and shape; ice or rock.

Glossary

Sound: Vibrations through a solid, liquid or gas. Some vibrations can be heard by the human ear.

Sound Energy: A form of energy that is made by vibrations.

Speed: How fast something moves over a distance.

States of Matter: Solid, liquid, gas.

Stream: A small flow of water.

Substance: Anything that has matter or material.

Sun: A big ball of gases which gives off light energy. The Earth revolves around the sun.

Survive: To remain alive.

Swamp: Low ground covered with water.

System: Parts working together to perform a task. These can be a life (circulatory), physical (lever) or earth system (volcano). Within systems there are transfers of energy.

Temperature: Measuring the heat energy of an object or substance.

Thermometer: A tool to measure heat energy or temperature.

Tool: An object that is used to do a special job or task.

Tropical: A climate around the equator where it is warm all year long.

Tundra: A treeless region near the Arctic with long winters and short cool summers.

Vapor: A gas.

Variable: Something that can be changed.

Vibration: To move back and forth quickly; quiver.

Vitamin: A substance needed by the body to remain healthy.

Volcano: A mountain made of lava, rocks, and ash that came from below the Earth's surface.

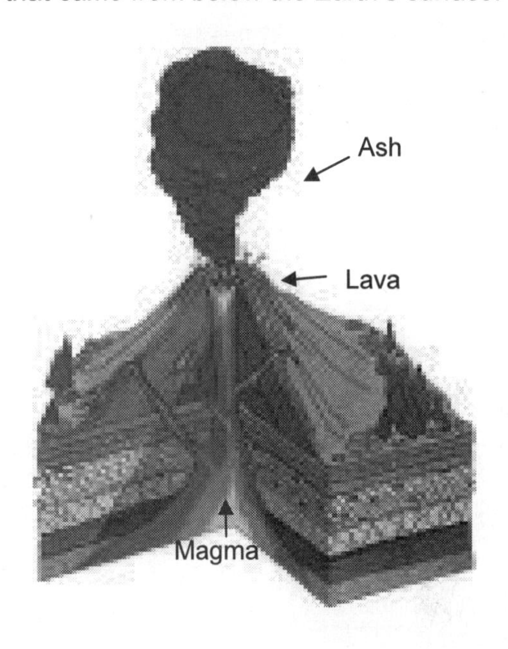

Volume: The loudness or quietness of a sound. OR: Amount of space taken up by an object or substance.

Water: A clear, colorless, liquid that is necessary for all forms of life.

Water cycle: The constant changing of water through the 3 states of matter that happens on and above the Earth's surface; (precipitation, condensation, evaporation).

Condensation: To change from a gas into a liquid.

Precipitation: When any form of water falls to the Earth in the form of rain, snow, sleet, and hail.

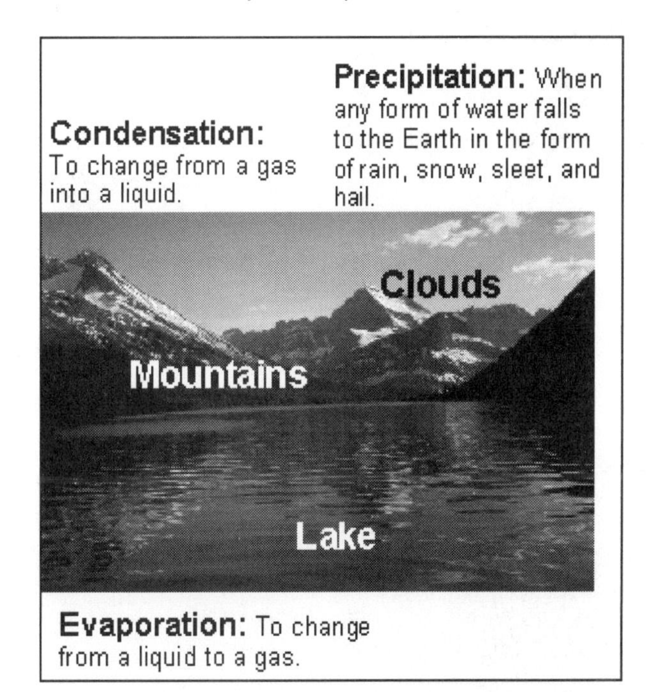

Evaporation: To change from a liquid to a gas.

Glossary

Water vapor: Water in a gas form.

Weather: The changing conditions in the atmosphere over a short period of time.

Weathering: The process of breaking down and wearing away rocks. This is not erosion.

Wedge: Two inclined planes put together; a simple machine like an axe.

Weeds: Plants that grow where they are not wanted.

Weight: The amount that gravity pulls on an object. Weight is different from the mass.

Wind: Moving air.

Lesson 1: "If it is to be, it is up to me."
Anonymous

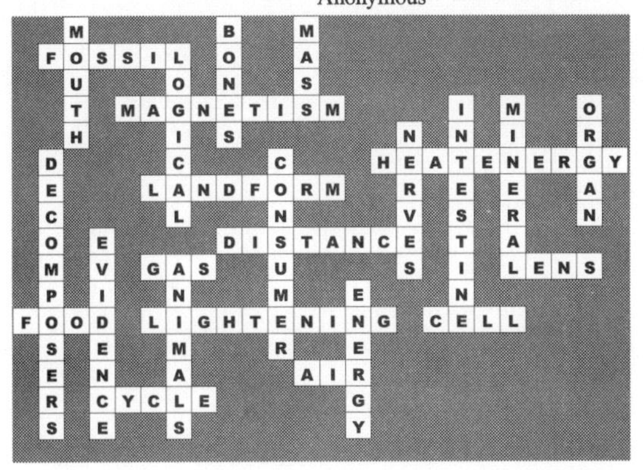

Lesson 2: "Good enough never is."
Debbi Field, Founder of Mrs. Field's Cookies

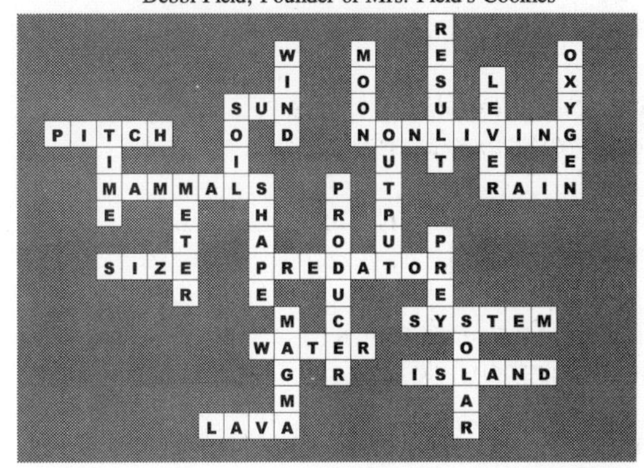

Lesson 3: "If you stand at all, stand tall."
King James I of England

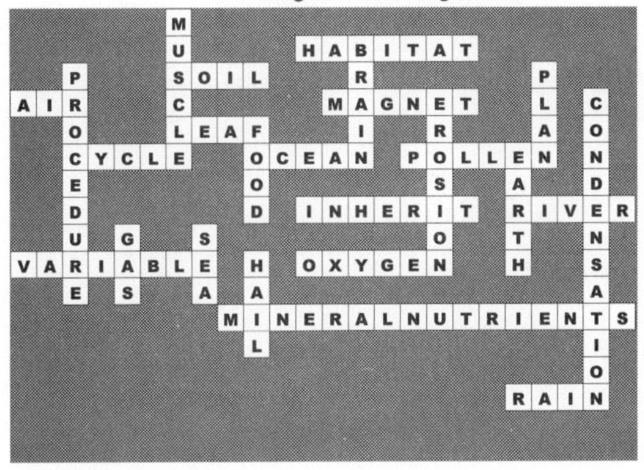

Lesson 4: "Problems are only opportunities in work clothes."
Henry Kaiser, Builder

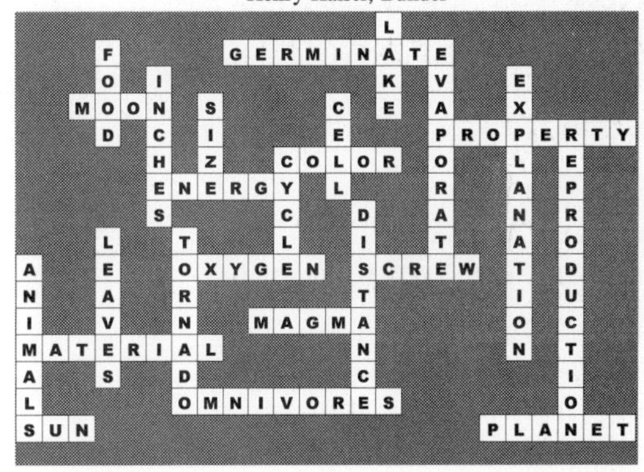

Lesson 6: "There's a better way to do it. Find it!"
Thomas Edison, Inventor

Lesson 7: "What would you attempt to do if you knew you could not fail?"
Dr. Robert Schuller, Minister

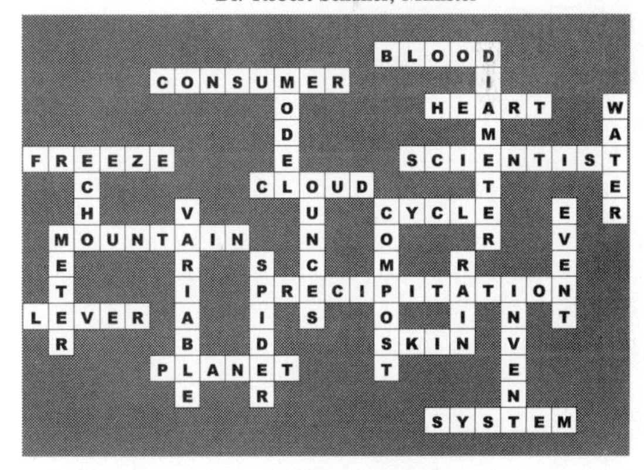

Lesson 8: "One of the secrets of life is to make stepping stones out of stumbling blocks."
Jack Penn, Author

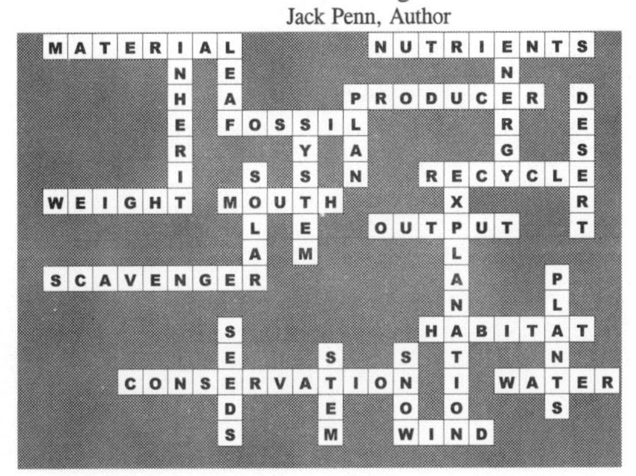

Lesson 9: "The manner in which it is given is worth more than the gift."
Pierre Corneille, Playwright

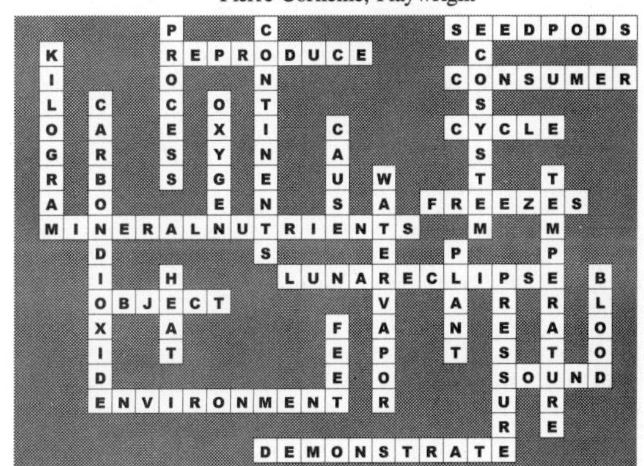

Lesson 11: "Light tomorrow with today!"
Elizabeth Barrett Browning, Author

Lesson 12: "The things that count most in life are usually the things that cannot be counted."
Bernard Meltzer, Attorney

Lesson 13: "The times change and we change with them."
John Owen, Poet

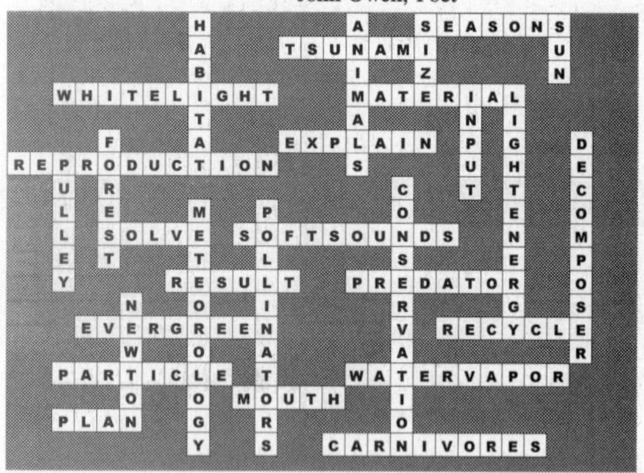

Lesson 14: "Let no one ever come to you without leaving better and happ
Mother Theresa, Missionary

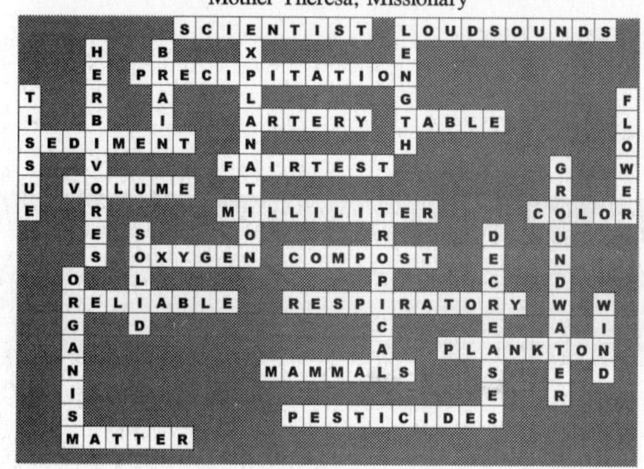

Lesson 16: "Advice: It's more fun to give than to receive."
Malcolm Forbes, Business Leader

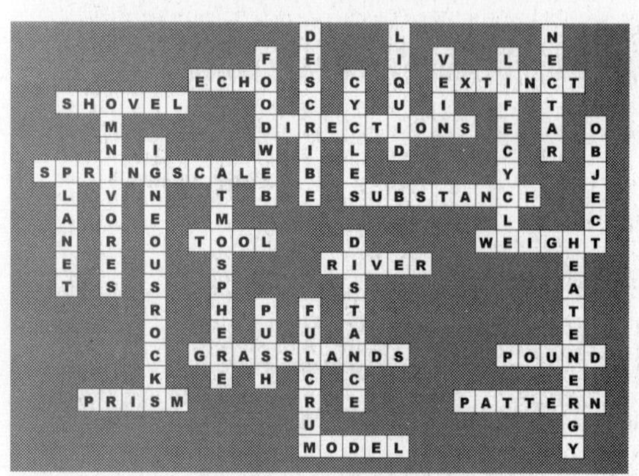

Lesson 17: "Well done is better than well said."
Benjamin Franklin, Statesman

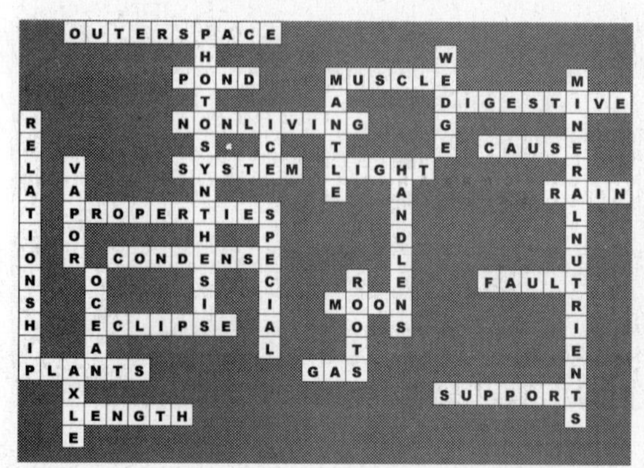

Lesson 18: "We make ourselves rich by making our wants few."
Henry David Thoreau, Philosopher

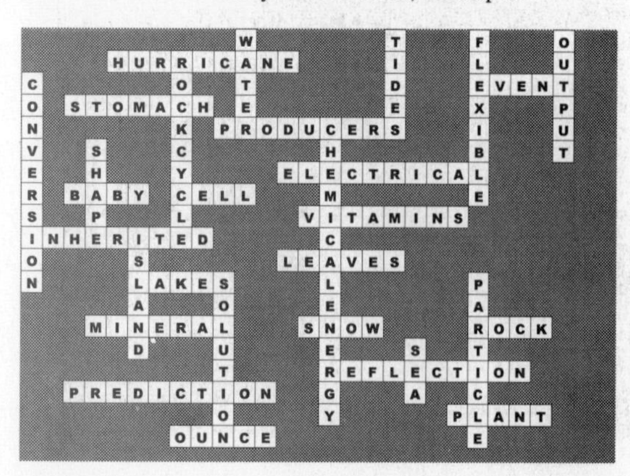

Lesson 19: "It is never too late to become what you might have been."
George Eliot, Author

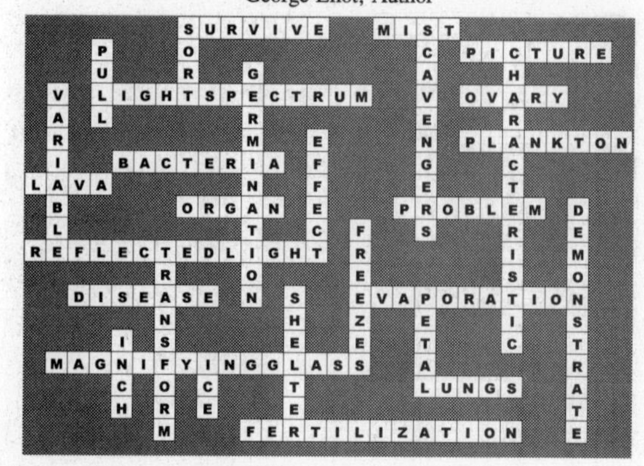

Lesson 21: "Before I speak, I have something important to say."
Groucho Marx, Comedian

Lesson 22: "A problem is a chance for you to do your best."
Duke Ellington, Musician

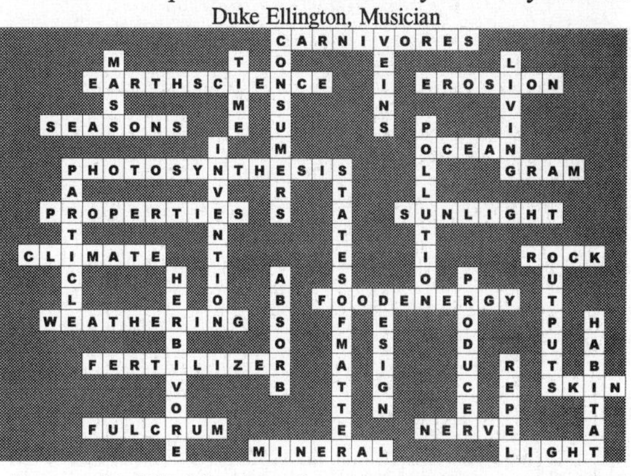

Lesson 23: "Obstacles are things a person sees when he takes his eyes off his goal."
E. Joseph Cossman, Salesman

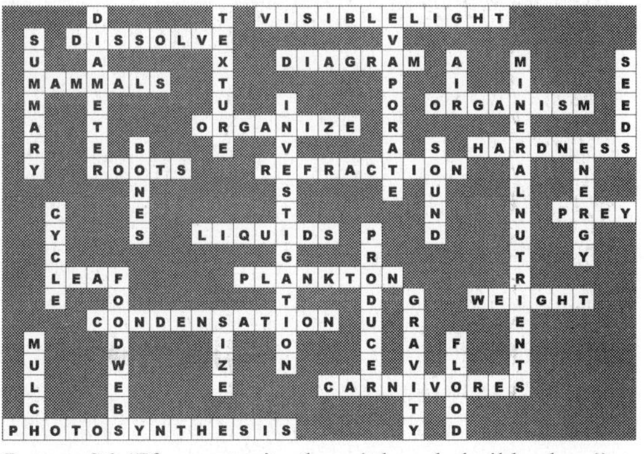

Lesson 24: "Nothing lasts forever, not even your troubles."
Arnold H. Glasgow, Psychologist

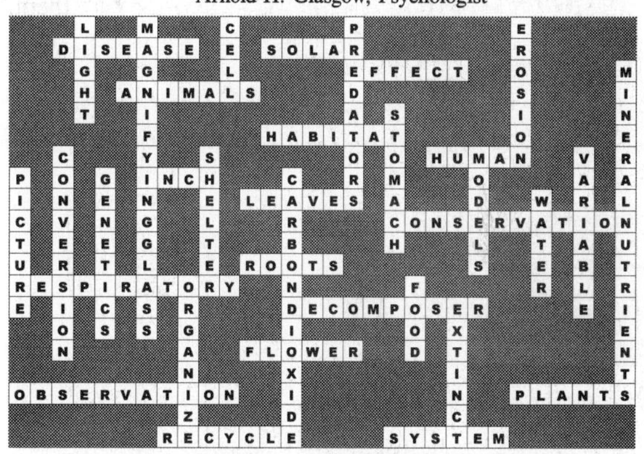

Lesson 26: "If opportunity doesn't knock, build a door."
Milton Berle, Comedian

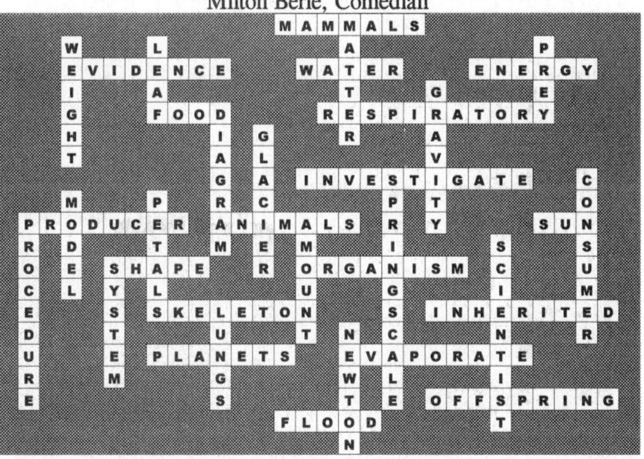

Lesson 27: "To be upset over what you don't have is to waste what you do have."
Ken S. Keyes, Jr., Author

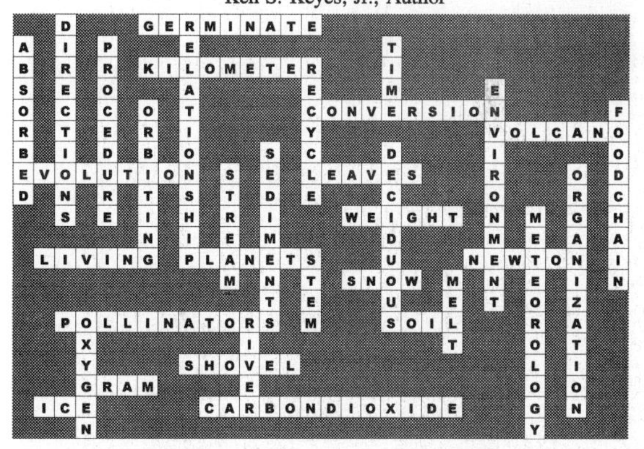

Lesson 28: "You have to have a dream so you can get up in the morning."
Billy Wilder, Actor

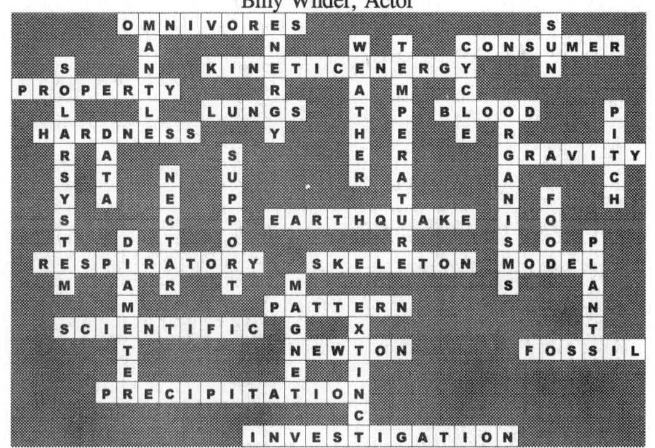

Lesson 29: "Even if you're on the right track, you'll get run over if you just sit there."
Will Rogers, Humorist

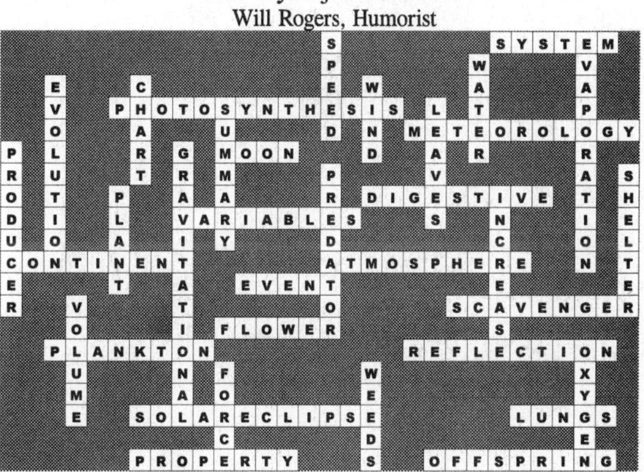

Lesson 31: "To create one's own world takes courage."
Georgia O'Keeffe, Artist

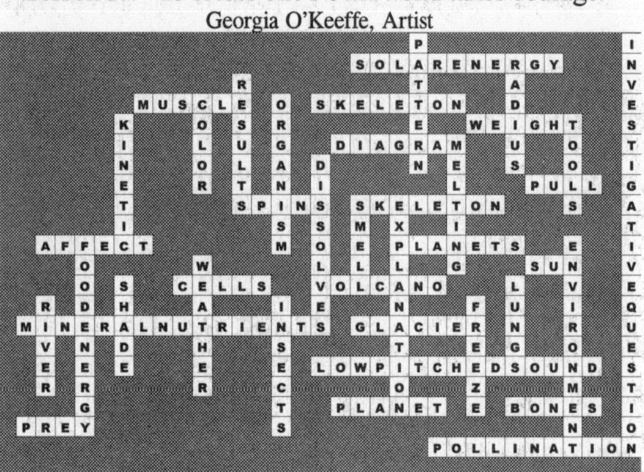

Lesson 32: "Imagination is more important than knowledge."
Einstein, Physicist

Lesson 33: "A friend is a present you give yourself."
Robert Louis Stevenson, Author

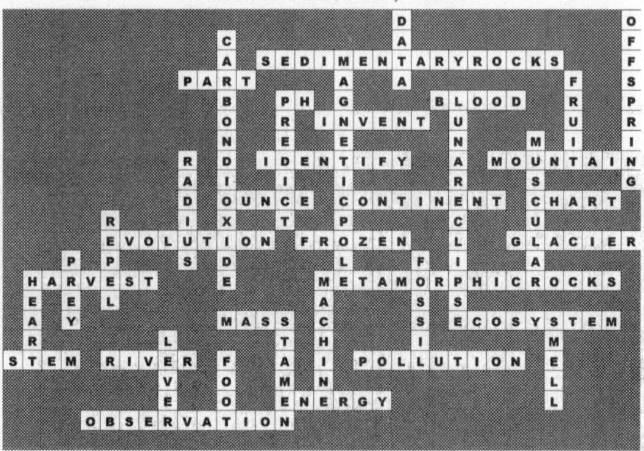

Lesson 34: "You learn something every day if you pay attention."
Ray Leblond, Author

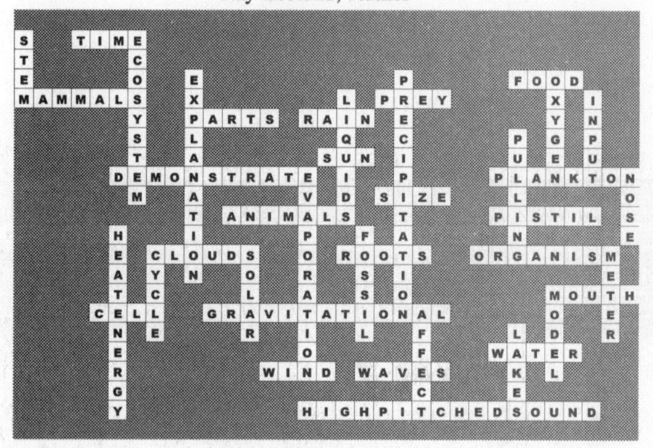

Lesson 36: "In order to achieve anything, you must be brave enough to fail."
Kirk Douglas, Actor

Lesson 37: "Never mistake motion for action."
Ernest Hemingway, Author

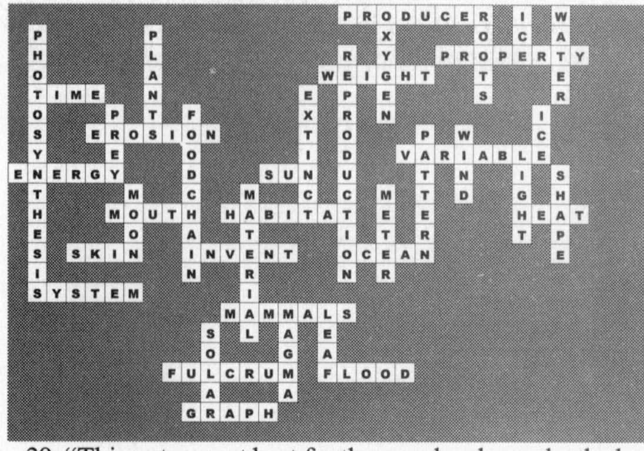

Lesson 38: "Whenever I don't have the answer to something, I find someone who does."
Walt Disney, Cartoonist

Lesson 39: "Things turn out best for the people who make the best of the way things turn out."
Art Linkletter, Personality

Lesson 41: "The fragrance always stays in the hand that gives the rose."
Hada Bejar, Author

Lesson 42: "A book is like a garden carried in the pocket."
Chinese proverb

Lesson 43: "We cannot become what we need to be by remaining what we are."
Max de Pree, Author

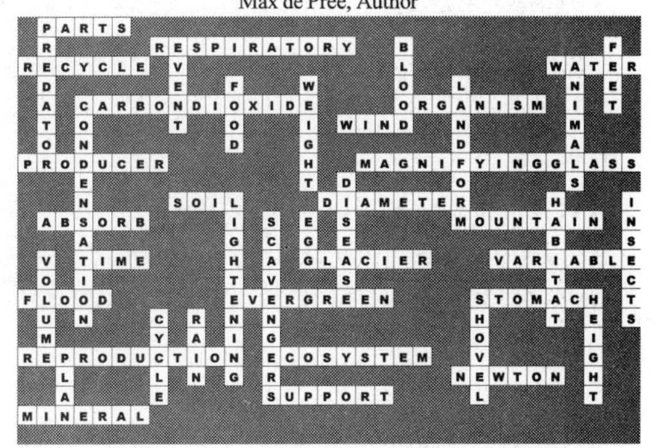

Lesson 44: "What the heart knows today the head will understand tomorrow.'
James Stephens, Author

Lesson Activity Answers

Lesson 1: 1. B 2. C	Lesson 2: 1. B 2. A	Lesson 3: 1. B 2. A	Lesson 4: 1. C 2. A
Lesson 6: 1. A 2. B	Lesson 7: 1. A 2. C	Lesson 8: 1. A 2. B	Lesson 9: 1. A 2. C
Lesson 11: 1. B 2. C	Lesson 12: 1. B 2. B	Lesson 13: 1. B 2. A	Lesson 14: 1. C 2. B
Lesson 16: 1. B 2. B	Lesson 17: 1. B 2. A	Lesson 18: 1. C 2. B	Lesson 19: 1. C 2. A
Lesson 21: 1. A 2. C	Lesson 22: 1. C 2. A	Lesson 23: 1. B 2. C	Lesson 24: 1. A 2. B
Lesson 26: 1. C 2. B	Lesson 27: 1. A 2. B	Lesson 28: 1. C 2. A	Lesson 29: 1. A 2. B
Lesson 31: 1. A 2. C	Lesson 32: 1. B 2. C	Lesson 33: 1. B 2. C	Lesson 34: 1. B 2. A
Lesson 36: 1. A 2. B	Lesson 37: 1. A 2. C	Lesson 38: 1. B 2. B	Lesson 39: 1. C 2. A
Lesson 41: 1. A 2. B	Lesson 42: 1. B 2. A	Lesson 43: 1. C 2. C	Lesson 44: 1. B 2. C

Suggested answers for **Short Answer Responses** are available online for some of the questions, though teachers may choose to use these activities as opportunities to explore new ideas and concepts in science. Suggested scoring rubrics are supplied at the back of this book to the "assessment activities." (Assessments are lessons 5, 10, 15, 20, 25, 30, 35, 40, & 45).

Lesson 5: Assessment 1

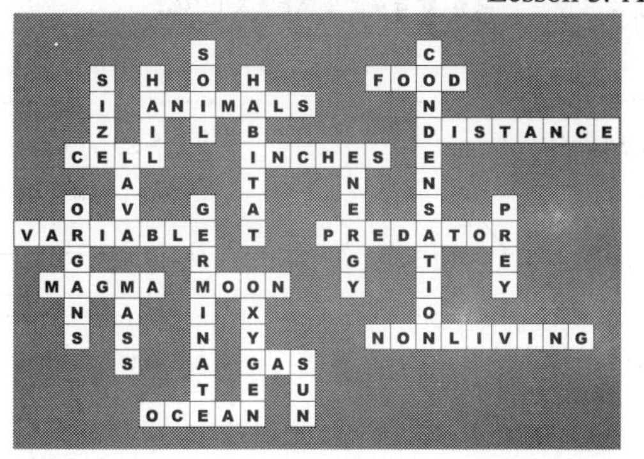

Lesson 5:

1. A 2. C

3. The student should draw and label a plant system to include roots, stem, leaves and flowers.

AND

Explain what would likely happen if one of the parts of a plant system were missing.

(**Examples**: Without flowers the plant could not be fertilized and couldn't reproduce. **OR** Without leaves the plant couldn't do photosynthesis and wouldn't be able to make its own food.)

Lesson 10: Assessment 2

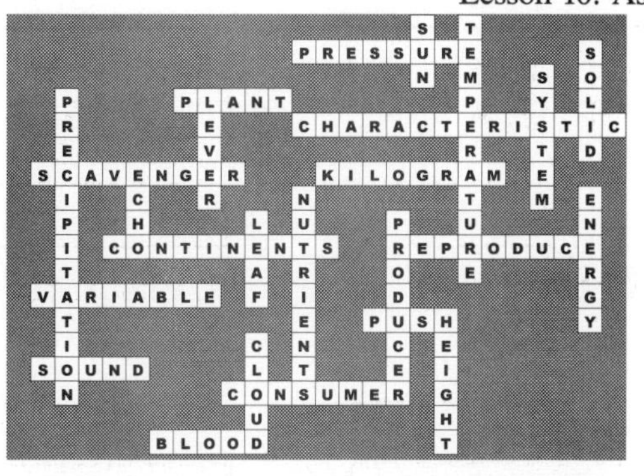

Lesson 10:

1. B 2. B

3: Identify **two** affects on the river system.

- Roots can't hold soil and water becomes muddy.
- Without shade the water gets warmer.
- Animals lose their habitats.

Explain how the salmon species could be affected.

- Mud could kill eggs.
- Temperature could kill off eggs.
- Salmon population increases due to less prey.

(There are many more reasonable responses.)

Lesson 15: Assessment 3

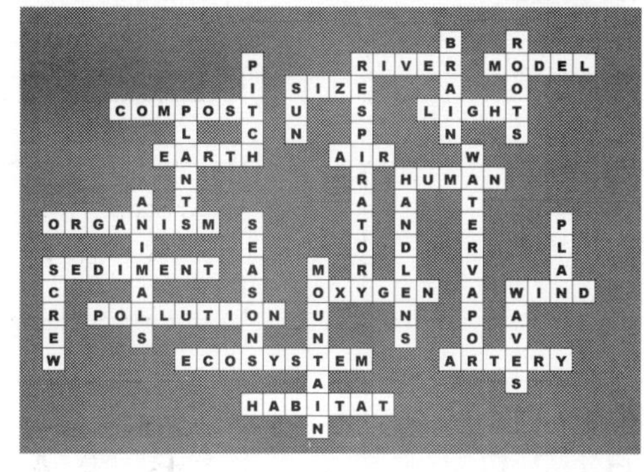

Lesson 15:

1. B 2. C

3: **Prediction** should refer to both the manipulated variable (temperature of golf ball) and the responding variable (height of bounce). Example: I think the colder (warmer) ball will make the ball bounce higher (not as high).

Procedures should include the manipulation of the temperature as the ONLY change. The student should make efforts to keep other variables as controlled as possible. Student should include measuring, recording, and repeated trials.

Lesson 20: Assessment 4

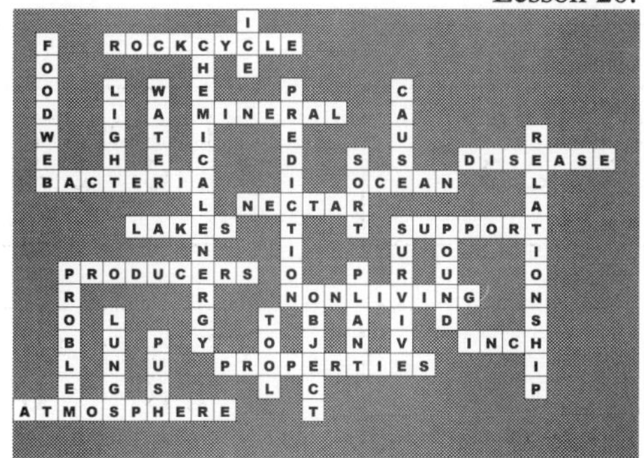

Lesson 20:

1. C 2. B

3: The student should identify **two** changes that could be made to the rocket system to improve flight distance.

- Size of balloon
- Length of straw
- Amount of tape

Explain how these changes would allow the rocket to fly farther.

- Heavier rocket has more momentum.
- Larger balloon holds more air to push longer.

Lesson 25: Assessment 5

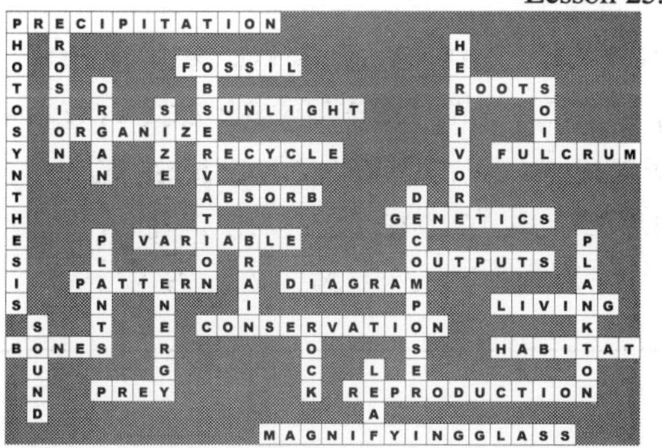

Lesson 25:

1. A 2. B

3: **Prediction** should refer to both the manipulated variable (amount of baking soda) and the responding variable (temperature of the solution). Example: I think adding more lemon juice will make the temperature higher.

Procedures should include the manipulation of the amount of baking soda as the ONLY change. The student should make efforts to keep other variables as controlled as possible. Student should include measuring, recording, and repeated trials.

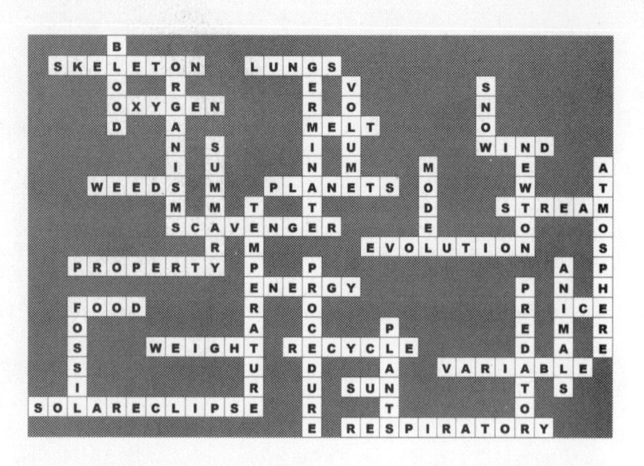

Lesson 30:

1. C 2. A

3: Student should state that the prediction was not correct. Include their results of high and low data averages in your explanation. (Take data from one column.)

- i.e. at 40 degrees: 93 seconds (high data)
- i.e. at 100 degrees: 30 seconds (low data)

Compare the results of high and low data averages.

- The lower temperature lasted much longer.

(A mathematical analysis of data would be great as well – 63 seconds longer)

Lesson 35: Assessment 7

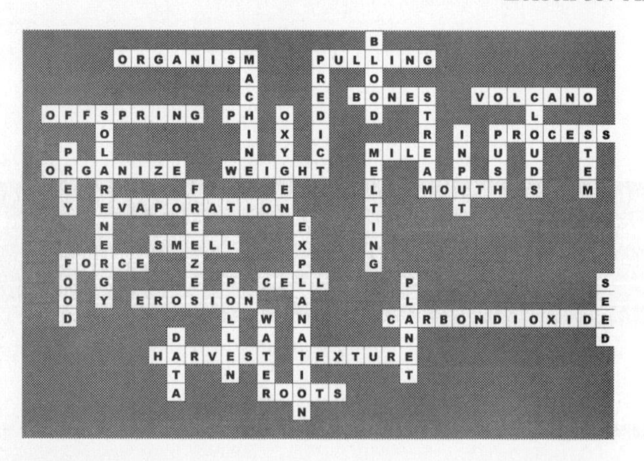

Lesson 35:

1. A 2. B 3.

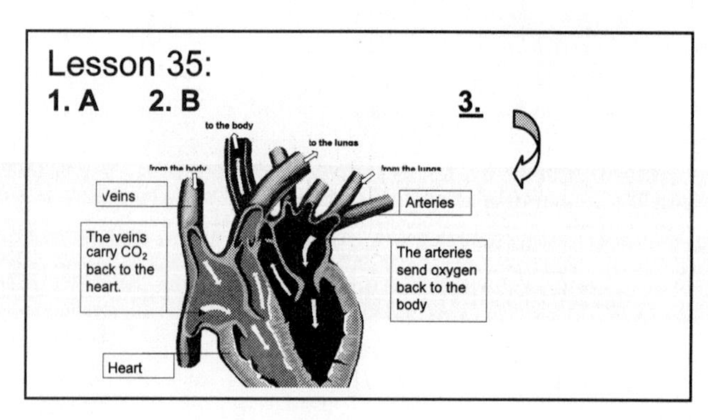

Lesson 40: Assessment 8

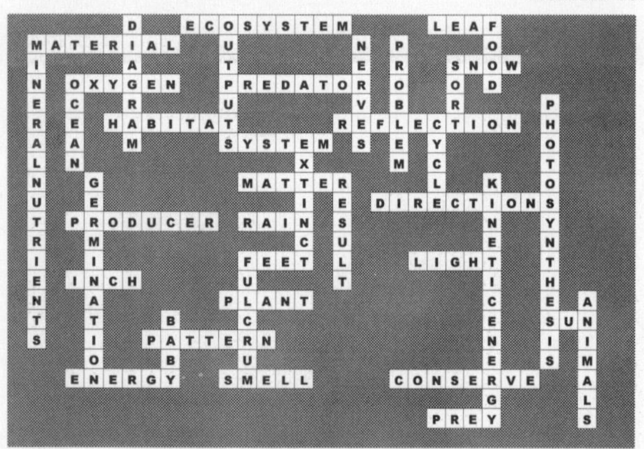

Lesson 40:

1. A 2. C

3. Identify **two** ways that organisms could affect the river.

- Beavers building dams.
- Moles or muskrats digging holes.
- Humans building structures.

Explain how these changes would affect the flow of water in the river system.

- Dams hold the water back or re-route the water.
- Holes change the natural direction of the river.

Lesson 45: Assessment 9

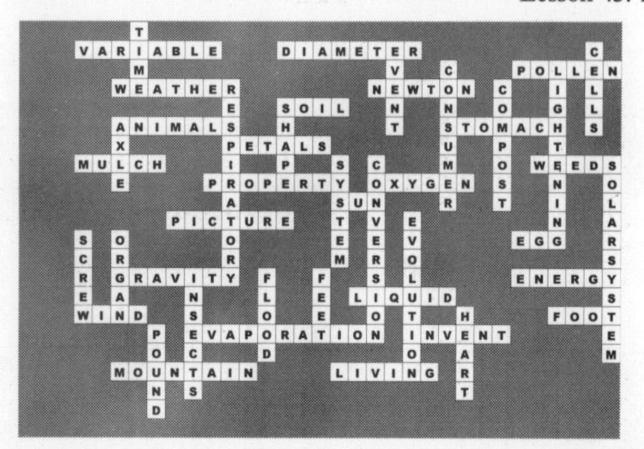

Lesson 45:

1. B 2. C

3. Identify **two** possible benefits of composting to the environment. Examples would include:

- Adding nutrients back into the soil
- Keeping landfills from over-filling
- Healthier soil = healthier plants = healthier animals
- Improved physical structure of the soil
- Used for mulch to improve erosion conditions, severe cold conditions or drought conditions